Tactical Lock Picking Vol. 1 - Foundations

by Pat Watson

WARNING

The author and publisher are not liable for any injuries, damages, or losses resulting from the use or misuse of the techniques described in this book.

Readers are strongly encouraged to seek professional training, and to exercise caution when using any sharp or dangerous tools.

Access only those locks for which you have been given explicit permission by the rightful owner, or when you have otherwise attained permission both morally and legally.

UNCENSORED TACTICAL

Website: UncensoredTactical.com
Shortcut: UTAC.io

ISBN: 978-1-7348254-4-2 (hardcover)
ISBN: 978-1-7348254-9-7 (paperback)
ISBN: 978-1-7348254-5-9 (ebook)

Ordering Information:

Special pricing and usage is available (with our prior written approval) for large orders, for reselling, for education purposes, and for families that would like to teach their homeschool children about the illusion of locked-security and the value of locked-security.

Disclaimers:

- We strive for accuracy in information but we are human. This book is intended to be a teaching tool, if you find that we have gotten something objectively incorrect, please write to us so that we can learn from you and better teach others. Email: Pat@utac.io

- We have always prided ourselves on teaching an "open curriculum", *you* should always be the arbiter of what you add to your personal tool box and what you don't.

- Try to have some patience and grace with yourself, others, and with us throughout our works. While we strive for professional and accurate instruction, we do not claim to have an encyclopedic knowledge of obscure lock trivia. Our goal is to create a guide for physical results first and foremost.

- You alone are responsible for your actions. Your source of learning and your tool and technique of choice is irrelevent. If you don't have the moral right to kick someone's door in without permission, you probably also don't have the right to pick their locks to gain access.

- Our branding is in fact "Uncensored Tactical" and while we reserve the right to occasionally use adult language and to make adult comments, we rarely engage in that these days; although it does occasionally happen so you are now warned.

- The frequent changes between the use of "I, we, me, us, UTAC, my, our, (etc.)" is not a mistaken writing point of view, it is because UTAC is owned and operated by the author alone, but I am blessed with a co-author who wishes to remain unnamed as well as a great community of collaborators and friends. Some statements and stories are unique to the author alone and some statements represent my co-author or my friends and peers etc.

- Thank you, sincerely, for trusting UTAC with your training time and your training budget. Please enjoy.

Dedication:

*There is more value to be learned from dissecting why your experienced a failure,
how you managed it in real time, and what you did after it,
than from simply documenting your successes.*

*Your failures are a gift. Go try. Go fuck up.
Then learn, and go get better.*

*See the previous page for complaints about our language.

Training Agenda

01 - "BROKEN BUT NOT DYING"
11 - MANUAL UNLOCK CHECK (X2)
13 - SKILL: "RAKING" TECHNIQUE
15 - WHAT TACTICAL LOCK PICKING "IS" AND "ISN'T"
29 - THREE LINES OF GEAR
53 - "SUICIDE BY PILLS"
59 - TWO-STEP RULE
67 - "LT'S OFFICE"
75 - SKILL: UNDER DOOR TOOL
79 - "LOCKED IN"
85 - SKILL: MASTER 175 BYPASS
89 - "EMBARRASSED"
93 - SKILL: COMBO LOCK FEELER-GAUGE DECODING
99 - SAME LOCK SAME KEY
101 - FOUR-DIGIT CODES
117 - "CANCELLED ENROUTE"
119 - "POOL FENCE REDEMPTION"
123 - LEGAL VS MORAL
125 - VEHICLE ENTRY CASE STUDIES
131 - GATES AND CHAINS
141 - SKILL: RESTRAINT ESCAPES
151 - TARGET ASSESSMENT
159 - THE FIELD IS NOT YOUR COUCH
169 - FAILURES
170 - DEFENDING AGAINST CMOE ATTACKS
171 - FINAL DEBRIEF & RESOURCES
173 - REVISIONS

CASE STUDY: BROKEN BUT NOT DYING

When it comes to lock picking, the social media comments will tell you "YoU dOn'T nEeD tO LeArN LoCk piCkiNg, jUsT uSe BoLt cUtTeRs oR a ShOtGuN!" Well, there's not really a good place on a door to use your bolt cutters, and I have a fairly educated guess that most of the people that are "pro-breaching shotgun" in the comments section probably don't know what standoff distances and angles are and they probably don't know what or why there are specialized industry approved "breaching rounds" but I digress.

If you ask people who use entry skills on the daily to save lives, the following case study is an unbelievably common example that I hear students bring up to me in my courses that "[This type of case study] is exactly why lock picking has a home in emergency response.

Enjoy.

"Can any unit advise if they have a lock pick set?" Dispatch called on the radio.

"Three-William-Sixty-One, I'm available for that call," I responded. I worked the West District. Dispatch explained that our East District was requesting a non-destructive entry to assist on a fire/medic call. In the local law enforcement agency I worked for, I was the only go-to "entry specialist" in our patrol division. This was in fact a self-earned and self-appointed title. I told dispatch, "Clear it with Sergeant 150 please, and attach me. I'll be 51 (enroute)."

My sergeant responded on the radio, "This is Sergeant 150. Approved."

As I pulled up to the scene, I saw several squad cars parked on the street near the suburban residence with a fire truck and a paramedic truck right up front, in the driveway. I had to park about a football field away, behind all the other emergency vehicles. Knowing ahead of time that I was there for a non-destructive entry, I grabbed my 2nd Line Gear bag from the trunk of my patrol car and started the hike toward the front door.

TARGET ASSESSMENT STEPS 1 & 2: MORAL RIGHT FOR ENTRY & LEVEL OF URGENCY (MRE/LOU)

I went straight to the front door and met with the patrol deputy, Deputy R, who was the Primary on scene (basically the same rank as every other officer but the first one on scene who becomes responsible for delegating tasks to everyone else). "Hey! You're here!" he said to me. "We've got an elderly woman inside. Lives alone with her dog. Slipped and fell. Immobile, probably a broken hip. She's begging these Fire department responders not to smash her door in because she's on fixed income. We need you to get us in with as minimal destruction as possible."

told him that because the home was a cookie-cutter residential home, there was a high probability of a quick and painless entry. With the information that a victim was locked in the structure who needed our help, Moral Right for Entry (MRE) was achieved; morally it was the right thing to do to make entry (destructive or non-destructive). But since a life was on the line, I sought out a medic to check the liability of my ever-changing Level of Urgency (LOU) requirement.

tapped the shoulder of a nearby medic. "Based on your information, since this is a medical call and not a law enforcement one, do you need us to smash a window and crawl through against the patient's request? Or do you feel comfortable letting me do a quick walkaround (Target Assessment) first and to try to respect her request to not damage anything?"

The medic and I had a short conversation concerning:

- The patient's ability to clearly communicate on the phone to the 911 dispatcher.
- The patient's ability to talk loudly enough from inside the home to be heard by on scene personnel outside.
- The evident loss of blood or consciousness.
- Our ability to go destructive if we notice any change in the patient's stability.

The lead medic on scene gave me a verbal confirmation: "Yeah, go ahead. She's broken but not dying." I could therefore take a few minutes to poke and pick around before we had to decide whether to break a door or smash a window.

turned to Deputy R again: "You are in fact the Primary here, right? Where is your supervisor, just so I can say I got a sergeant's as well as a medical approval?"

For the professional use of this skill set, if an approval from a supervisor is wishy-washy and is anything less than a clear "yes" or "sure, go ahead," don't proceed. Or at least reassess your options.

We called for the East Side sergeant, and he met us at the front door. "Hey, Sarge, I can almost certainly pick my way into this home. I just want to keep you in the loop and get clear verbal approval. Fire says they're not in a rush and they're comfortable waiting until I do an assessment to try to make non-destructive entry." He approved. I was specifically waiting to hear the word yes or something very close to it. I made a mental note to make sure to document his approval in my write up of my actions on scene, after the call.

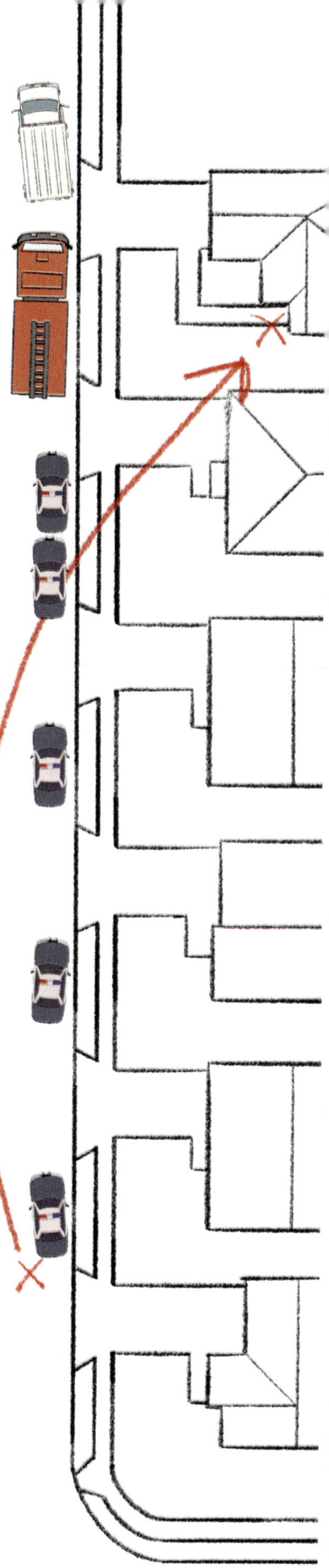

TARGET ASSESSMENT STEP 3: ATTACK VECTOR IDENTIFICATION

With the Moral Right for Entry and the Level of Urgency (MRE/LOU) figured out, it was now time to continue down the path of my decision-making process into the actual entry phase.

In about two seconds I assessed the front door by identifying all of the Attack Vectors that were known and available to me:

1. No visible hinges on the outside of the door because it was a push door (you turn the handle and push the door to open it, or "activate the thumb press" on this particular door).
2. A deadbolt was on the right side of the door, but not visible. Because this was a push door, the frame occluded our vision of the latches. This means I can't "walk the deadbolt" like you sometimes can on a pull door.
3. A window was just around the corner to the right of the door. If we had to go destructive, we could at least go minimally destructive and smash a small window and reach through and manually unlock the door. Even if the window was too far from the deadbolt I could still probably use a reach tool to unlock the deadbolt through said broken window.
4. For the front door, no bypasses were necessary or available, the keyway for the deadbolt was my only real attack vector so far.

My ballpark guess was that it would take up to 10 minutes to rake open the front deadbolt (at the time of this entry, I didn't spend a ton of time practicing a more direct picking attack known as Single Pin Picking [SPP]. I mainly focused my skill set on quick raking techniques and several bypass options).

TARGET ASSESSMENT STEP 4: MANUAL UNLOCK CHECK (X2)

Step 1 and 2 are almost automatic in the field. You'll almost always have the chance to instantaneously confirm whether you have the right to enter a locked building and how quickly you need to get inside. While these two steps are mostly automatic, it's still important to consider them to make sure you start on the right track.

Step 3 can be as simple as a quick glance to determine your entry points, or your Attack Vectors. For trickier obstacles like multiple padlocks on a chain wrapped around a gate, you may want to spend a few minutes maneuvering the chain around and getting a better perspective.

Step 4 is often your first physical act on scene. If the obstacle is a doorknob, try the knob, take a deep breath, then try again harder. If it's a padlock, push and pull the shackle to see if it pops open, take a deep breath, then check again harder. (This sequence will be repeated ad nauseam in this book and for good reason!)

Looking at the front door to the residence, I grabbed the handle and pressed down on the thumb activator and then gave the door a hard push and pull. I took a deep breath and tried again, putting my shoulder into the door to make sure it was locked. It was locked. Step 4 - Manual Unlock Check - complete for this door.

Another way to confirm that a deadbolt is activated (extended in the locked position) on a push door (or even sometimes on a pull door with a very tight door to frame fitment) is to slide a flexible shim into the door jamb above the deadbolt area, then slide the shim down to the level of the deadbolt, and see if the shim makes contact with the deadbolt or if it can slide down past where the deadbolt should be.

TARGET ASSESSMENT STEP 5: TOOL AVAILABILITY

For a nondestructive entry on this door, I had to do a keyway attack. Because I was called to this scene specifically for an entry, I was already prepared to immediately start with tools from Second Line Gear bag. Some of which are shown below:

- Multiple types of rakes, picks, and tension wrenches
- Bump keys
- Pick gun (& Electric Pick Gun - Similar functions)
- (Residential Pin-Tumbler Lishi tools - not yet released to the market at the time that this case study took place)
- (Minimally destructive "through the lock")
- (Destructive door kick or window smash)

[Above: $25 Stainless Steel Bogota pick set.]

[Above: $40 Bump Hammer + a Bump Key from a 5pc set.]

[Above: $50 SparrowsLockPicks.com "Snap Gun".]

[Above: $175 Southord Electric Pick Gun.]

TARGET ASSESSMENT STEP 6: "NOW OR LATER?"

Knowing I had approval to take my time, I decided to walk around the house and look for an entry just screaming to me, "I'm easy! Choose me!" instead of burning time and trying to pick a lock that could take (on a bad day) 10 minutes or more. I, therefore, chose to start the entry attempt "Later" for Step 6.

Prior to scouting out the next obstacle to assess, I figured I'd press pause real quick in order to make the most of having so many people on scene.

Economics plays a big part in Tactical Lock Picking (TLP), specifically the economics of time management. One of the basic skills you'll develop during your Target Assessment is knowing whether to start picking the first lock you see or to do a thorough inspection and determine the range of possibilities before starting.

TARGET ASSESSMENT STEP 7: OTHER RESOURCES & RESET BREATH

Since I had several pairs of hands, I figured I'd make use of them as "entry resources". "Hey, Primary, you know how to pick a lock? Ever tried it?" The primary officer on scene told me he owns a set of bump keys at home and understands the concept but that's about it. "Great!" I reached into my bag and found a set of common bump keys and a bump hammer and handed them to him. "If you know the concept already, then use these. I'm going to check the perimeter for another option for entry." Second-Line Gear bag in hand, I started off around the house.

Additional things to consider in your Other Resources step might include hidden keys or nearby passcodes, people on scene with skill sets who could assist you, a locksmith you could call while you work on other obstacles, a tool you could manufacture (or "makeshift") from your environment, a nearby friend or family member who might have a key to the structure, and a search using your smartphone for knowledge about any of the specific locked obstacles and Attack Vectors you've identified.

After implementing an "Other Resources" option to get more hands involved, I continued my inspection of the house.

I found a back door that had only one deadbolt and a passageway knob, or simple knob (a knob that operates a latch but with no locking mechanism and no keyway on either side - a basic turn-to-open-and-that's-it knob). The deadbolt face and keyway were visible but the actual latch mechanism was not, meaning I couldn't attempt to walk it, or jimmy it, to the side. The keyway, though, looked inexpensive, old, and worn. All things that *might* work in my favor.

Sometimes old worn and weathered locks are easier to pick and bypass, and sometimes they are harder. Even at advanced practitioner levels this can be hard to determine by first inspection and can only be determined once you are hands-on for your entry attempt.

I got on the radio and told the Deputy R, still at the front door, that I was attempting entry at a door in the back of the house.

Again I attempted a Manual Unlock Check (Target Assessment Step 4), I tried the handle and pushed and pulled on the door, took a deep breath, and tried again harder. I confirmed the door was locked. (This concept should be stuck in your head by the end of this book so that on scene, and under stress, these words will run through your thought process. Even professionals mess up this step in real life. Attempt access, breathe, then attempt again harder. Push, pull, up, down, twist left, twist right, etc.)

Now for Target Assessment Step 5 again (Tool Availability). At this entry point I knew I had all of the same tool options as before (minus one set of bump keys). I therefore zoomed to Target Assessment Step 6 "Now or Later" and decided to try to pick this lock "now". At this point we have switched from the assessment phase and into the physical entry attempt phase.

I took out a $25 tool set from my wallet (a.k.a. My First-Line-Gear). This set consisted of two pieces: (1) a Bogota rake with a bend in the back to use as a tension wrench and (2) a single hump (half diamond, single peak, etc.) also with a bend in the back that could be used as a tension wrench. (This is the first make/model lock picking set I ever used, years and years ago, and we now sell them in our online web store. They are not only some of the least expensive lock picking kits around, and some of the smallest makes/models around, but they are also one of the most effective and most widely used professional sets around.)

I spent about two to three minutes on a "raking" type attack (randomly trying to get the pins inside the keyway to line up correctly). The sergeant walked around the house to the back and asked me for a progress report. "I'm real close, Sarge," I said. "Any second now and I think I've got it."

There are two common picking methods that most lock pickers start with, Single Pin Picking (SPP) and raking. With SPP, you apply tension to the keyway and "feel" each pin individually and then lift each pin to its correct height (this is sometimes depicted in video games with a somewhat close-to-accurate detail). The raking method is most effective on low-security locks. Because of the mass production of locks and the inexpensive machining process and very loose tolerances in the product, you can exploit the lock's design by applying light turning tension on the keyway and randomly raking/sweeping/jumbling (scientific terms) the pins around until the lock opens.

At this time in my life, I was able to overcome a lot of obstacles with the raking technique. I didn't spend a ton of time practicing SPP, but based on a lot of experience I knew this lock was a great candidate for a simple raking attack. I was confident I needed just a little more time.

The sergeant, rightfully, was getting impatient and keyed the mic on his radio, "If 3W61 can't get this lock open in the next two minutes, I'm gonna call it and we'll just have Fire use their breach..."

Click.

[Above: Lock at rest, deadbolt extended to the left, all the way into the door frame in the locked position.]

[Above: Lock keyway partially spun, deadbolt slightly receeded but, still not clear of the door frame.]

"Got it!" I yelled to the sergeant, who was then standing just around the corner of the house from me.

With one audible click, the plug started to rotate (the plug being the small cylinder into which you put a key). But due to poor fitment of the door, **the keyway stopped spinning short of unlocking the deadbolt**. Why? Because so many doors are installed/fitted improperly. At some point in your life, in order to lock or unlock a common deadbolt, you probably had to apply a little pushing or pulling force on the door to assist you. Sound familiar? That's what was happening here. The extra pressure on the deadbolt latch made it difficult to fully spin the keyway.

From my Second-Line Gear bag I used pliers to grab the flexible tension wrench close to the lock to try to apply extra force without bending it too badly, but the amount of force needed bent the tension wrench. Then I searched for a stiffer tension wrench. Because this wrench was a little more rigid, I was able to use a lot more force to get the deadbolt to start to turn.

With some metal-on-metal screeching, the obstacle unlocked. Then, after a twist of the handle, the door easily pushed open.

CONCLUSION

The sweet old lady's lap dog was quickly upon us, barking at our ankles. The medics were able to extract the patient out through the now unlocked front door. No damage was made to the patient's home and we were able to lock the doors upon our exit; *something that destructive entry is unable to offer.*

Many responders get hyper focused on the "application of services" during an emergency but they often forget that in the course of providing services you may leave the subject of those services open to unnecessary further injury to their person or property when you treat them like a number instead of treating them like a whole human. With this skill set, **options exist** to not have to leave a patient's house wide open with a smashed window or a kicked in and now unlockable front door.

Lessons were learned, a new case study was recorded, and our Tactical Lock Picking curriculum was honed a little bit more (as should happen on every entry).

- I bought a selection of much more rigid tension wrenches to help overcome future stubborn locks.
- I also could have used any number of generic tools in the keyway like a screwdriver or another small pry tool, noted.
- I did not remember to take photos of the keyways but it was also a medical call so I did not want to detract from services being provided. I did not approach the patient to ask for photo permission. The photos in this chapter are recreations of similar locks and doors.
- I'm glad that I did remember to use my human resources on scene to have two people attempting entry at the same time instead of being the only one working on a possible entry.
- I would call the informing of my supervision and the double-checking with Fire/Medics a win that should repeatedly occur during a professional setting.
- Having a Second-Line Gear bag pre-packed was a big help, and initially bringing it to the scene, instead of going back to my car for it, was a smart choice. This is the difference though between "Being called to assist with an entry" and "Realizing in the moment that you are locked out and need to make entry."

Note: At the time of this case study, and at the time of the writing of the first edition of this book, residential Kwikset and Schlage "Lishi" tools were not yet introduced to the market. They are now a fantastic addition to almost anyone's Second Line Gear bag. If I had an appropriate Lishi in my gear bag at the time, there's a chance I may have used one on the front door immediately and sent other Deputies on a perimeter check to find easier entry options, rather than myself. I would have directed them to take photos with their smart phones of the doors and locks that they find and to bring the photos back to me if I still hadn't gained entry yet.

When used correctly they can very rapidly achieve entry. They're not magic, and sometimes they require a few minutes of struggle and trouble-shooting but more often than not they provide very quick access to residential pin-tumbler keyways.

[Left: A $75 KW1 model "Lishi 2-in-1 Decoder" tool being used in a Kwikset clone (Gatehouse-brand) deadbolt.]

The troubleshooting:

1. Once I was unable to use the current tension wrench to finish completely spinning the keyway (plug), I grabbed pliers to try to apply some extra force. But using pliers wasn't enough. The tension wrench was too flexible and the force from the pliers ended up bending and damaging the current tension tool.

[Above: Due to pressure on the latch from bad door fitment, the semi-flexible Bogota tool was unable to maintain rigidity past this point.]

[Right: With a stiffer tool, the deadbolt was finally able to scrape past the threashold and into the fully unlocked position.]

2. I switched to a slightly stiffer tension tool and used pliers to assist with the last bit of screeching sliding metal-on-metal, receding the deadbolt the last bit of the way into the door and thus into the fully unlocked position.

CURRICULUM:
MANUAL UNLOCK CHECK (X2)

"Consistency". In the military and law enforcement "SWAT-type" breaching world, and in the Firefighter "forcible entry" world, they both have a try-before-you-pry rule. This is built into their entry sciences as an intentional step because even highly trained and competent professionals can overlook an already unlocked door.

It's just a part of life: have hammer in hand = use hammer to solve nail-problems. It's easy to get tunnel vision while running up to a door with large breaching tools in hand, ready to apply said tools while overlooking simply trying the knob.

The same thing exists in the lock picking world. Before excitedly shoving your lock picking tools into a keyway, try the handle.

"Try the knob, take a deep breath, then try the knob again harder.
Twist left, twist right. Push, pull. Up, down. Etc."

In the professional entry fields (SWAT, Fire, and others too) the method that is often demonstrated, taught, and carried out in thousands upon thousands of test runs has nestled itself into what is known as a "training scar", reinforcing an action in training for the sake of practice/drills/etc. that ends up being carried out in the field as a detrement to successful operations. This method is where operators (enforcement or fire, etc.) will walk up to a door with tools in hand, and they will KNOW that the purpose of this drill is to USE their breaching tools, so they grab the knob with only their thumb and forefinger and they give the knob an every so slight jiggle left and right, resulting in the "play" in the handle rocking back and forth about three to four degrees left and right. The problem is that in the real world there are hundreds of thousands of knobs that have that same three to four degrees of play in the knob before the slack is taken up to activate the latch even when the knob is unlocked.

Be consistent. If you are about to apply large or small entry tools, always try the handle first. And instead of trying that little thumb and forefinger tickle, try it the same way you would try *your own front door handle*:

"Commmmeee onnnn! Open up! Squeeze! Turn left really hard! Turn right really hard! Come on you stupid door!"

...obviously there are exceptions like "surprise" and the safety of the entry operators to consider. But, double checking in a consistent and diligent manner is a step to shore up a gap in common field failures, and it has earned a place in our curricum.

[Opposite: If you routinely wear gloves in your professional capacity, or even just due to your environment, but you have never tried picking locks with gloves on, *you're gonna have a bad time*!] :)

SKILL: "RAKING" TECHNIQUE

Step 1:

- **Find a good-fitting tension wrench (you'll learn fitment as you progress) and insert it into the bottom of the keyway.**
- **If "barely touching the wrench" is 1%, and bending it to where it breaks or deforms is 100%, you want about 10% of that pressure.**
- **The fit of the tension wrench, the amount of tension you provide, when and at what speed you alter your tension by increasing or decreasing the amount ... THAT is the magic here. Any primate can apply step 2 on the next page and still achieve an entry as long as the tension is applied correctly.**

Step 2:

- **Choose your "rake" tool and insert it into the keyway above the tension tool so that the rake will be touching the pins that are inside the keyway. To check, look inside the keyway with your tools removed in order to determine which side of the keyway the pins are on (top or bottom).**
- **Move the rake tool in and out of the keyway and up and down (like a tooth brush). The rake should never have to "twist" clockwise or counter-clockwise while you are attempting to unlock the keyway. Just straight in, straight out, and up and down flatly, or up and down at an angle like a see-saw.**
- **This "raking" motion is way less important than the type and amount of tension that you provide.**

That's it! Just explore these two variables "Tension" and "Pin Manipulation" for a while and you should be able to get your first ever lock picking entry as long as you're using the correct types of tools on the correct type of locking mechanism.

Tip. You should never be using so much force that you start breaking your tension tools or your rakes/picks. There are exceptions but, I've been doing this for close to two decades now and I break less than one pick a year doing raking techniques.

WHAT TACTICAL LOCK PICKING "IS" AND "ISN'T"

In modern psychology there's a maxim that goes something like, "It's very difficult to understand what normal behavior is. It is much easier to look at what is not normal and to go from there."

Because so many people ask me if I'm a locksmith when I say that I teach lock picking, and because a lot of people who research lock picking end up in hobbyist "locksport" circles, and because lock picking also has a special home in the hacker and security testing communities, and because lock picking is often portrayed as a super secret spy skill on the big screen for movies and television, we will start with what Tactical Lock Picking (TLP) isn't and we will go from there.

LOCKSMITHING

Locksmiths will often have some combination of a brick and mortar storefront and a work vehicle to drive to job sites. It was a surprising fact for me to learn that locksmith businesses do not all offer the same suite of services across all of the lock-spectrum but that they commonly specialize in only a few areas of expertise like residential installations, ID card systems, vehicle lock repairs or lockouts, commercial door installations and door hardware, etc. Although almost all locksmiths are required to be adept at the installation, repair, replacement, and upgrading of at least some type of hardware - not to mention the logistics and administration of the business side of operations (forms, paperwork, receipts, inventory tracking, etc.).

The services that locksmiths provide are 100% at the convenience of the business. Yes, it is often in a company's interest to complete work quickly and to be available for additional jobs, but the company has the absolute right to turn down or to postpone any type of service whenever they choose.

While their projects may vary, locksmiths have the ability, in combination with their storefront, their vehicle, and a tool belt/bag, to store and transport lots of expensive, large, heavy, specialty, and duplicate items to a job site.

Depending on their specialization, locksmiths might each own a few dozen different decoder tools for different makes and models of vehicles (as well as backups and duplicates) and associated manuals and training material, which would be relatively expensive. Or they might keep hundreds of key blanks in their vehicle along with a few different large and bulky key-cutting machines.

We do have multiple courses that can turn an operator into someone that can pull information from a key or a lock and then generate working keys for said lock in the field, but turning the team vehicle into a locksmith van would be unnecessary and would logistically be a non-starter.

A locksmith would also be expected to store all the tools and devices required to remove, fix, and install security hardware: drills, hammers, pry bars, arrays of different screwdrivers, boxes of nails and screws, and tapes and glues that could add up to a lot of weight and a lot of space, let alone the cost.

While it may be in the interest of locksmiths to complete some jobs in a hurry, they often have the luxury of time to select and organize the gear they need, to choose the time of day, to wait for weather to clear, to set up proper lighting, to retrieve different tools, to review manuals, videos, or resources, etc. through their own business lens and decision making matrix.

At times locksmiths do have to open or access a specific lock, and at other times they don't have to access a specific entry point but are free to be creative. For example: if a piece of a key is lodged into a deadbolt keyway, the locksmith must address the problem with that deadbolt, but when a locksmith has to gain access to a residential home, they may be free to access any different number of locked obstacles.

There are things that locksmiths don't specialize in, like some of the higher-end destructive entries. It's probably a rare case for a locksmith to show up to someone's front door and say, "No problem, we'll have you in there in a jiffy," and then to go grab a battering ram from their locksmith van. Probably not appropriate for the vast majority of locksmith calls.

LOCKSPORT

Just like there are competitive sports, competitive shooting competitions, competitive butter sculpting and competitive beard grooming, there is a growing global community of competitive lock picking practitioners and this community and their competitions are known as "Locksport". Also, nothing wrong with competitive butter sculpting or beard grooming. Forget normal, go be unique, go be weird!

Locksport is a rapidly growing field of practice but it is still not a household name and, when people do hear about it, it is still something that is easily misunderstood by people who have never picked a lock themselves.

Generally speaking, the Locksport community focuses on using picking tools and tension wrenches to open up difficult pin tumbler (and similar) locks and sometimes in timed trials, by expertly manipulating the pins inside the lock with great finesse and fine motor skills. Members of Locksport often practice their craft on flat desk surfaces, with locks mounted in a vice, in perfect lighting, in comfy chairs, with whatever specialized tools they have decided to lay out on the tabletop in front of them.

Some members do practice offshoots like safe-cracking competitions and other lock entry feats besides just pin-tumbler locks. But what I almost never see in Locksport circles is bypassing.

While bypass techniques are often known in the Locksport community, these techniques are rarely "competitive." Sticking a bypass tool into a lock and twisting the lock open (an act that completely bypasses the challenge of solving a problem with finesse and practice and hard earned knowledge and experience) wouldn't make for much of a competition. A bypass technique in Locksport competitions would be equivalent to driving backwards from the start on a racetrack and saying that you've technically gotten to the finish line first.

Locksport is the sport or recreation of defeating locking systems. Its enthusiasts learn a variety of skills including lock picking, lock bumping, and a variety of other skills traditionally known only to locksmiths and other security professionals. Locksport followers enjoy the challenge and excitement of learning to defeat all forms of locks, and often gather together in sport groups to share knowledge, exchange ideas, and participate in a variety of recreational activities and contests. - https://en.m/wikipedia.org/wiki/Locksport (2019)

A question that is frequently asked about Locksport is what the legality is. One of the most common questions I get about any type of lock picking tool or any type of practice of the craft is "Yeah, but isn't that illegal?" The answer is, "Most likely, no. Probably not illegal." Toool.us (and Toool.org; yes there are three o's on purpose) has a very helpful running list of lock picking laws by jurisdiction. We will cover more on legalities later in this book.

Having spent over a decade of time between federal law enforcement, military operations, and local law enforcement, I can tell you that most (yes, overwhelmingly "most") police officers aren't familiar with lock-picking laws. The sport of Locksport, the act of lock picking, the tools of the craft, and the legality of it all, are becoming more publicly known but are still widely misunderstood.

Locksport offers a mental and physical challenge, requires great skill and dedication, and is a lot of fun, not to mention addicting, for sure. Locksport is a great start to building your understanding of locks and to exercising your fine-motor skills, but Locksport alone will not turn you into a Tactical Lock Picker.

HOBBYISTS

While many in the Locksport community may practice Locksport "as a hobby," it wouldn't do to not mention the pure hobbyist.

While there is definitely a broad spectrum of hobbyists (which sometimes includes the Locksport community), there are those who are content to sit on their couches with a $25 lock pick set and the same old three to four cheap padlocks, just picking away while binge watching their favorite TV show. And this is totally fine! I support this!

Others who fit into this "hobbyist" category are those who enjoy the puzzles or who have stumbled on lock entry techniques and have stayed curious, continue to test their skills, and are open to exploring and learning related skills.

Mindset is what places people into this category - the mindset of approaching locked challenges casually, as occasional puzzles. If lock picking were an iceberg, hobbyists would be the ones that are happy with just the tip.

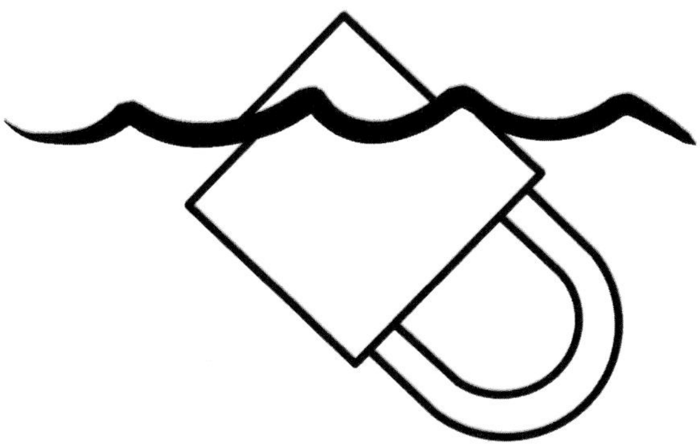

RED TEAMS/PENTESTERS/OTHER

While there is a debate in the security industry about the difference between a Red Team and a Pentesting Firm, or other similar designations, we are going to start with a definition from the Red Team podcast.

> A real Red Team is a group of highly skilled professionals that mimic and simulate a real adversary. This team will research, investigate, learn, and become the adversarys they are trying to portray. They will also research the organization or thing they are trying to red team, understanding its culture, the industry, or area they are located at and what makes them run.
>
> By consciously working to assume the adversary's role, a "thinking enemy," a Red Team can provide a realistic view of how a new idea, plan, or approach can perform in the real world, helping leaders to understand and address the risks in every aspect of the business, and providing unorthodox views on problems and their solutions. - https://redteams.net/about (2019)

Very generally speaking, security assessment professionals spend much less time with lock picking than with other security assessment tasks such as accessing computer networks, taking advantage of stairwells and elevator access systems, cloning ID key cards, drafting (sneaking) in through doors behind employees, digging through trash, and even using disguises and props like clipboards and name tags to gain entry into places. These security assessment professionals also spend quite a bit of time in administrative and logistical roles with marketing, attending meetings and briefings, and performing other bureaucratic functions.

Lock picking does have its role in this industry, but you'll have a hard time getting hired as a "pentester" if you don't know how to turn on a modem or a router.

TACTICAL BREACHERS/POLICE, FIRE, ETC.

There are individuals in law enforcement, firefighting, and other services who know how to pick a simple padlock, but the topic is rarely taught within agencies. I've attended some very expensive "restricted" government training courses and at most, lock picking was a single bullet point on a single slide. There are exceptions, but the content found in this book is little-known in most military and law enforcement communities.

Tactical breachers often have a selection of core tools that are very big and bulky and are used in very specific scenarios and that require specialized knowledge and training. Notice I said "knowledge and training" are required. While in some cases if you hit a door hard enough with a sledgehammer, you may get it to open, there are many more cases where if you have the perfect tool and the perfect setup but are untrained you may completely exclude yourself from making a successful entry. Examples of "perfect tool, perfectly paired obstacle, untrained operator, resulting in a failed entry" could easily fill an entire book.

Here is a brief example (and a common one): Drunk person comes home and is locked out of their house. They watch a lot of TV and they assume that you can easily kick in any door with no training required. They take a step back from the door, they lean forward with a shuffle-step start, they kick the doorknob hard, they twist their ankle, they fall, and the door doesn't budge. (For the record, no. We do not aim for doorknobs as our targets even when we have decided that a door kick-in is the right choice for our mission requirements.)

Tactical breachers often have entry authority before they even set foot on a property, and they often (but not always) have an increased Level of Urgency to make entry, regardless of the consequences of property destruction. This is especially true for Firefighters in response to visible fires taking place. The destruction of an entire door and its frame is much preferred to the burning down of an entire building. Some people in the Fire community teach less-destructive, or semi-destructive, techniques (often referred to as "through the lock"). But most of the techniques that Firefighters employ place human life and fire control way above any destruction of property on the scale of concern.

Because of the situations that professional first responders routinely face, they are uniquely responsible to be familiar with widely destructive entry methods (cutting human-size holes through walls, etc.). If we are only looking through the scope of "life over property" then all bets are off, smashing windows and destroying doors fit well into this equation. (There are times, however, that non-destructive techniques are way faster, and safer, than almost any destructive technique!)

WHAT TACTICAL LOCK PICKING "IS"

The systemized approach of using lock picking and similar non-destructive methods to overcome locked obstacles in emergencies or during operations.

One of the biggest goals of TLP, and the structuring of it, is to deliver the best possible return on investment for the practitioners of this craft. It's about the energy and resources you put in, being able to give you the maximum amount of immediate field effectiveness.

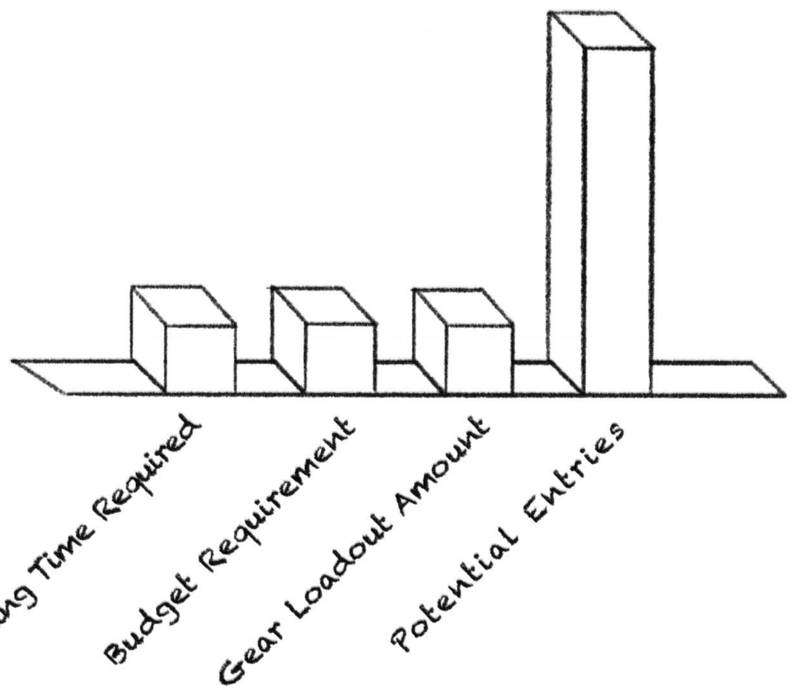

KEY FACTOR 1: THE FRAMING

Built into the UTAC Tactical Lock Picking curriculum are two steps prior to attempting any entry, which obligate you to determine (1) whether you have the moral right to make entry and (2) how much of a hurry you are in. Written into the curriculum, as well as taught and reinforced during case studies presented in the classroom, is the principle that students make a habit of determining whether what they're doing is morally right.

Also part of the framing of this skill set is that it is to be used in emergencies to overcome locked obstacles. Emergencies can range from being late to work to experiencing large-scale physical disasters. We'll cover more on this later but an emergency lockout is way different than planning to make covert entry into a target facility at a future point in time. Because "overcoming locked obstacles during emergencies" is our primary skill set focus, we do not focus on the repairs, replacements, installations of locks and (at least in our TLP courses) we don't have to worry about how to rekey or repin locks, etc.

UTAC's book "Tactical Lock Picking - Vol 2." is our curriculum on how to generate working keys in the field for practical and tactical applications. That curriculum DOES leverage things like repinning and rekeying locks and compliments our foundational TLP curriculum very nicely.

[Opposite: View from the inside of the EnergeticEntry.com breaching-training facility. A door designed spedifically for practicing mechanical breaching.]

Additionally, the framing of our TLP curriculum is designed to be inclusive of creative options. While the curriculum is a great baseline to start with, feel free to, in the words of the late great Bruce Lee "Absorb what is useful, discard what is useless, and add what is uniquely your own" (and variations of this quote). I don't limit students by saying, "We cannot use this technique because it is not a part of our curriculum!" While TLP focuses on non-destructive entries, we have never intended to replace mechanical, ballistic, or explosive breaching. Our curriculum is designed to be added to the toolbox of any professional or prepared person as additional capabilities. You can always call yourself a practitioner of Tactical Lock Picking, even if you happen to be a Firefighter who routinely responds to visible fires and rips doors off of their hinges 99% of the time.

UTAC does offer a Covert Entry course that uses lock picking and bypassing and similar skills to help plan for future operations, and that curriculum also pairs well with the systemized approach of TLP, but it is specifically designed for people who are doing recon and planning ahead to make entry into a building. All of the TLP approaches can be used to assist in that venture but there are additional decision making matrices to consider when planning ahead for a covert entry.

Tactical Lock Picking is framed for people who find themselves locked out during emergencies and have to make use of limited resources. They get to choose between their Every Day Carry entry tools on their person (First Line Gear), the entry tools they have stored in a go-bag somewhere (Second Line Gear), and their assortment of larger, bulkier, heavier, and other items that don't have a permanent place in the first two lines of gear (Third Line Gear). Fourth Line Gear includes resources, references, research, and other types of informational considerations that can assist with your non-destructive entries during an emergency.

[Photo below: view from the inside (loced side) of a door, watching a flexible latch shim tool manipulate a door latch for entry.]

KEY FACTOR 2: THE TOOLS

When forming this curriculum, the following questions called for an answer: If you're in a hurry for an entry, wouldn't it be better if you had the right tool with you instead of having to go get it or to wait for someone to bring it to you? Wouldn't it help if you knew precisely which tools you had on hand and which were in the trunk of your car or in a nearby go-bag? Wouldn't it be better if your tools were stored in a way that was easy to organize and easy to carry? It might look cool to have a big plastic bin in your trunk stuffed with tools and lock-opening toys, but wouldn't you look a little silly during an emergency returning from your vehicle with an armful of random gear cradled in your arms, trying to keep items from falling out of your grasp?

Some tools are easily accessible to the general public and others are more specialized, more expensive, and sometimes the sale of some non-destructive entry items are regulated. Many tools are actually available in your environment if you know what to look for. And many tools have multiple purposes and others do not.

I have considered all this to help guide people towards which tools to start with to serve as a baseline. I designed my Three Lines of Gear to organize how to select tools, store them, and even how and when to deploy them in the field. This is discussed in our "Guideline: Three Lines of Gear" chapter.

Economics always applies (and not just for money). Even with an unlimited budget, even if you could technically purchase every lock picking and locksmith tool on the market:

- The training time just doesn't exist for you to be 100% perfect with every single tool and technique.
- You can't carry every single tool in your pockets, on your belt, in a bag, or in your vehicle, or even in a fairly large office.
- You can't bring every single tool out into the field for operations.
- You can't have a perfect maintenance history on every item.
- Even if you could do all of the above, you'd still have to know how to make the right choices in order to make the most effective and efficient use on scene while selecting from all of those available items.

[Opposite: A "Luggage Tag" flexible door latch shim in the process of slipping open a door latch from the "push side" of the door.]

KEY FACTOR 3: TARGET ASSESSMENT

Equipped with multiple tools and multiple techniques and confronting multiple entry points, it's in our best interest to have an order of operations that makes logical sense and that functions as a constant reassessment between the most likely and the quickest methods of entry (not always a balance that is easy to identify).

Our Target Assessment phase helps entry specialists understand what they're doing, when they're doing it, and what they should have done before, during, and after their entry attempts. Target Assessment is a process that guides practitioners in how to make quick calculations on scene (or even before arriving on scene) in many different ways.

In an emergency, one of the foremost factors in how we choose to access locked obstacles is time. If someone is actively dying on the other side of a door and you wait too long to open that door to provide help, their chances of perishing increase with every wasted second. We talked about this in the previous chapter. Even as the patient in the last chapter was deemed to have a level of consciousness that was acceptable, that could have changed at any moment, which would have immediately changed our type of entry and response options with it. The Tactical Lock Picker is a master at weighing and managing timeline expectations.

Using a Target Assessment system helps you decide how much of an initial assessment you should do once you arrive on scene, when to start the physical task of entry and on which attack vector, how to multitask and make the best use of additional resources, and how to determine the length of time that you'll try a specific technique before reassessing and considering other options.

KEY FACTOR 4: OTHER RESOURCES

An issue vital to first responders and prepared individuals is identifying and utilizing additional resources. Resources I've utilized in emergencies include tools delivered to me, maintenance personnel or family members contacted for spare keys, people on scene to assist me in my entry efforts, and locksmiths or even Firefighters called to scenes to assist with entries.

I've noticed that sometimes first responders will use a tool until failure and then move on to a different approach until failure and only after that will they call for a locksmith or a keyholder. Big mistake.

When a life is on the line, a much better use of time is to get all the resources you might need to start making their way to the scene. If you gain access before those other resources arrive, you can always cancel them. You can't always bring someone back to life because you chose to wait to call additional resources. In other words "I'm going to start picking this lock. Can you go ahead and start calling known family members and tell them to drive here with a key, now? And have any unit with destructive entry tools head this way as well, in case I can't get this lock open."

[Opposite: the UTAC "Enlightening Door" logo on our glow in the dark patches. Seriously, do ***not*** wear these during nighttime covert operations.]

Nothing says "tactical" like glow in the dark patches on a covert night time op!

(Humor, of course)
;)

KEY FACTOR 5: THE HUMAN FACTOR

In this TLP curriculum, there is a block that we have designed that takes the social habits of humans into account...sort of.

In the book Social Engineering: The Art of Human Hacking, author Chris Hadnagy defines his version of social engineering as "the art, or better yet, science, of skillfully maneuvering human beings to take action in some aspect of their lives."

In this UTAC system, we don't maneuver other human beings into taking action, but we do try to understand how people commonly interact with their physical-security environment.

Some common examples of how we creatures of habit commonly interact with our security systems: leaving certain doors unlocked, leaving padlocks hanging on chains that only appear to be locked, choosing shockingly simple and common four-digit codes, installing push button combination locks but never changing the factory-installed default code, using the same digital and/or mechanical four-digit code for every door in and around a facility, hiding keys and passcodes near locked the obstacles that they open.

And there are plenty more examples to point to where we don't engineer a human action to help us gain entry but where we de-engineer common human actions to gain insight into how to better overcome locked obstacles. In TLP, we maneuver ourselves into detecting and decoding how people may have interacted with their physical locked environment and we use that insight to help us get through locked obstacles.

This is our version of social engineering.

KEY FACTOR 6: SELF-SUPPORTED OPERATORS

If the UTAC team were asked to develop a policy manual laying out rules, regulations, guidelines, and policy for a "Tactical Lock Picking" qualification for a top tier military unit, we can and we would happily do so but, it would not be the same curriculum and the guidelines in this book. And before assumptions are made, let us help you understand this current curriculum by explaining why an operational team curriculum would look different.

The current UTAC curriculum for a Tactical Lock Picking standard is based upon the self-supported operator. We worked very hard to frame our curriculum guidelines so that anybody could add this skill set into their lives at a professional or even at an amateur level. Self-Supported Operators are people who have to literally "build this skillset into their lives" themselves, without the help of a huge budget or a huge bureaucracy.

Even police officers and many military operators can be considered a "self-supported operator". They may have their training budgets provided for them at someone else's expense, their ammo provided, their uniforms provided, their transportation provided, but they still may have to pay their own money to attend a lock picking course, and may have to still buy their own lock picking tools, and they may decided to stack their own library full of lock picking books. So, in one particular skill set, they have to "sell-support" even though they are learning and maintaining a skill for the purpose of field operations.

If we were to design a curriculum or a manual for use by an operator on a top tier team then we would be "designing a manual for someone with budget X, skill level Y, and resources Z". For the self-supported operator they only have their own individual budget, we assume a zero-skill level from the start, and we assume that you only have yourself as a resource and anything outside of that during an entry is a bonus.

Our curriculum can absolutely be implemented into a deployable operational team setting by scaling up. Super easy. If, however, we designed a military operational curriculum and marketed to the public "A Special Forces Skill Set - Made Available for YOU!" then we'd be lying. Because the general public "you" doesn't have the same budget, infrastructure, training time available, etc. as most top tier military units. This is not a value-judgment to say "you (general public) are not as good as special operators", this is a curriculum design decision that was intended to go heavy in the areas that are always applicable, like the Target Assessment and our systemized approach, and to leave other areas wide open for customization like "You should design your own Three Lines of Gear, here's a recommendation to start". Thus, our curriculum is the result of starting at the ground level and also being infinitely scalable with increases in budget and resources.

We talk about this self-supported operator concept quite a bit in our podcast (Uncensored Tactical - podcast) and it certainly is mirrored in our teaching style and in our curriculum.

Consider all the things you may be trained to do, equipped to do, and expected to do for your profession, and consider the things you had to learn yourself, and purchase yourself, even if the things are used in the performance of your professional duties. *Those* are skill sets that are "self-supported".

KEY FACTOR 7: FAILURES

Imagine a lock picker gets called to help unlock a front door in an emergency. They bring some standard hook and rake picks and a few tension wrenches with them to the scene. They wave off any attempt at calling for additional resources and they say, "Don't worry. I'll be able to get this keyway unlocked in a jiffy!" After 60 seconds of trying they start to get nervous. Then comes the excuses: "This keyway is a little rusty inside, and it's a little cold out, and they must have installed the lock wrong." Then after about three more minutes they start overcompensating and using way too much pressure on their tension tool because they're embarrassed and frustrated, causing fine motor skills to start to falter. Now they have a decision to make: how long do they continue trying and failing and how long before they tuck their tail between their legs and ask the other people on scene to make the call for other resources?

I don't want you to be that guy or that gal.

Failures are not to be shrugged off in this artform/system. They are to be welcomed, directly analyzed, and learned from. Being aware of the vulnerabilities of locks, the limitations of lock picking, the inability to see inside of a lock, and the presence of unseen hardware will help you to determine which tools to use, which entry points to choose, and what to do next if none of these initial attempts at entry are successful specifically because of failures that you can expect to happen.

The TLP curriculum is based not only on what has worked in the field but also on what has not worked and why, and how to avoid such failures.

I hope that during your training you struggle and you fail often. You can only learn one data point when something succeeds perfectly as attempted on the first try, but you can learn many data points when you struggle and fail. I also hope that when failure happens, you open your mind, you admit your failure, and you allow yourself and others to learn from it so that you can take this skill set to new heights.

CONCLUSION

Unlike locksmiths: Tactical Lock Pickers cannot always wear specialized tool belts, carry bulging tool bags, and store expensive and specialized heavy equipment in our vehicles. We don't have storefronts with racks of products at our disposal. Also, we must be able to act in emergencies and cannot always delay a project. We might need to attempt entry in the dark, the cold, or the rain. And we might have only one or just a few tools to choose from. However, because we are not locksmiths, in exigent circumstances we might be able to make certain types of destructive entry. Also, we don't need to worry about the knowledge or infrastructure required for installing, uninstalling, or repairing hardware.

Unlock locksport: Our locks won't be mounted in a vice on a table. We might have to reach through a gate, in the dark, to access a keyway facing downward, without being able to see the lock (true story). And unlike Locksport, a huge portion of our entry (and escape) techniques are based on shortcuts, hacks, and bypasses. We're interested only in efficiency and likely effectiveness, and any and all techniques that we're able to access on scene are available to us. While TLP starts with a core curriculum, it is not an exclusive set of skills. Students are encouraged to absorb what is useful for their personal version of this skill set and to discard what is not.

Unlike hobbyists: As many students of mine have confirmed after taking our courses, "The field is not your couch. A sub-one minute opening at home is often a 10-minute opening in the field with the same lock."

Unlike security assessment professionals: We are not in the business of hacking into network security or forging employee ID cards. But like security assessment professionals, we are in the business of researching patterns of human behavior and the workings of physical security. TLP does not include any tactics on disguising ourselves to gain entry or on drafting in through doorways behind employees. If we had someone with access to the structures in question during an emergency, we'd ask them for entry in order to help us save a life within! TLP is framed for emergency entry so the tools/tactics/techniques are presented in a way where covertness is not required. Security assessment operators almost always carry the requiredment to be covert. The security assessment field is a great example of a wealth of very interesting and valid entry concepts but they apply in very different framing settings.

Unlike tactical breachers: When a professional operating as a Firefighter or under the team member title "tactical breacher" gets called to a scene, they often already know their MRE and LOU to be "Legally and/or morally authorized; go right now." In TLP we frequently must figure out our Level of Urgency on the fly and determine on scene whether we can legally or morally make entry, and we must constantly reassess. Also, we don't always have the authority to perform wildly destructive entries - we don't routinely deploy with an axe in one hand and a Halligan bar in the other. When necessary, we do use basic mechanical breaching tools like sledge hammers, pry bars, or professional glass punches.

There are a small but growing number of places both online and in person where you can learn lock picking, but even some who teach super-duper "Law Enforcement and Military Only" courses don't truly teach the field application of this skill set.

As of the writing of this book there are very few courses that we would steer people towards besides our own. Some are terrible, some are valuable but for a different framing of entry studies, and a very few courses/instructors exist who we are willing to collaborate with and to promote and support.

Our goal in our foundational Tactical Lock Picking course is to take students that have zero experience in ever having picked open a lock, and to turn them into prepared field operators that are making successful complex field entries before the course is complete on the second day. We have also intentionally designed the curriculum so that people that *do* have experience in non-destructive entries don't get held back in class. This is because students in our courses can always sharpen their decision making regardless of how well they can pick open certian types of locks.

Some sources teach some of these things, but I am still shocked every time I teach a block of instruction and a student tells me that they've been to multiple entry courses (even courses specific to lock picking) and have never even heard of the content in one of our instruction blocks. Makes my little heart happy to hear that.

Noticing a lack of available training in the real-life application of all facets of this skill set, I saw a void and was determined to shore up this gap of wildly valuable information for those out there that need it.

[Picking a lock mounted to an obstacle is an entirely different game from practicing on unmounted locks.]

CURRICULUM: THREE LINES OF GEAR

FIRST LINE GEAR

First Line Gear is defined as "the tools you routinely carry on your person." Many people would call this their Every Day Carry "EDC." The most common places that could be considered First Line would be what's in your pockets and on your belt every time you leave the house or every time you gear up for work. For me, I carry all of my lock picking tools in my wallet and on my key ring. First Line can also include your belt (on duty or off) for things like a multi-tool, a flashlight etc. If you routinely wear something like a battle belt during your daily operations, and if you had a small pouch with entry tools on that battle belt, it would certainly be considered First Line Gear.

First Line does not include backpacks, satchels, purses, etc.

There is a variable here that we call 1.5 line gear (one point five line). This is something like a fanny pack (an extremely useful modular tool storage system, and chicks dig it), or a clipboard with storage in it, or it may mean a small purse or something similar that isn't necessarily "attached" or "in your pockets" but it's definitely not a backpack and it is carried more often and more easily than what you would call a "go-bag". If you sometimes wear hard plate body armor and on that armor you have a sizable pouch with entry tools, then it would be 1.5 Line when it's not being worn and it would become First Line Gear once you don the armor and it's attached to you.

It is very important for us to point out to you that these "Lines of Gear" are not rules. They are not places that you MUST store tools. You can put the same tool in one line of gear and put a duplicate in another line as well if that works for you. You may be a carpenter or a maintenance worker and it might make a lot of sense for you to use your literal tool box as your entry tool storage container.

What matters is that if you want to be prepared to make "emergency non-destructive entry" then "in an emergency" you should have some type of First Line tools on your person when you leave the house.

First Line Gear: In your pockets and on your belt. "Attached" to you.
1.5 Line Gear: Sometimes attached to you, less than a go-bag.
Second Line Gear: A "go-bag". Shoulder strap(s) recommended.
Third Line Gear: Doesn't fit into the first two lines, vehicle trunk or closet etc.
Fourth Line Gear: Information.

First Line Gear: In your pockets and on your belt. "Attached" to you.

1.5 Line Gear: Sometimes attached to you, less than a go-bag.

Second Line Gear: A "go-bag". Shoulder strap(s) recommended.

Third Line Gear: Doesn't fit in the first two. Stored in vehicle / closet etc.

Fourth Line Gear: Information.

[Opposite: The current iteration of UTAC's turn-key "First Line Gear Kit". Available at the **UTAC.io web store**.]

Multi pass
(Door Latch "Jim")
Luggage Tag
(Door Latch Shim)
S.S. Bogotas
(A "gold standard" set)
Bogota Pi set (3.14" long)
Hair Clip
"Covert Jigglers"
(For Wafer Locks)
4-pin Comb/2x EZ Decoders

- **Bogota Pi Set**. $45. Available at the UncensoredTactical.com web store and other retailers. Each tool is 3.14 inches long (π pi. It's a math joke!) These tools have a smooth rounded polished feel and not too bad of a handle to grip on to for being EDC-sized, basically credit card sized. Not everything, but almost everything that is a variation of the Bogota style design, is a pretty safe bet for quality. In my personal experience the only Bogota tool yet that I haven't liked is the "Springs" set. If you're just starting out in lock picking, do not buy a 300-piece set from Amazon. Just go buy almost any Bogota set to get yourself started. The Bogota line is not only affordable but also trusted by people who are at the top of their entry craft!

- **Stainless Steel Bogotas**. $25. Available at UTAC's web store as well as many other fine retailers. Probably "the" best starter set in the game, and at a very reasonable price point. The Pi set above has a "flat" tension wrench that goes into the keyway in-line with the vertical keyway. This S.S. set (often called "Original Bogota's") has a "bent" tension wrench function at the rear end of each pick, which means that they insert into the keyway and bend 90 degrees perpendicular to the keyway. With these two sets you now have four raking profiles, two lifter picks, two tension wrenches, and you can effect tension in all four of the cardinal directions (N, S, W, E), or on the face of the clock (12, 3, 6, and 9 o'clock). I would never design a selection of tools for EDC without being able to at least hit the flat and bent tension wrench options.

- **Luggage Tag**. $8. Available at OscarDelta.co.uk and other retailers. If you're starting your entry studies journey, and you're starting to build your lock picking loadout, once you have some type of lock picking tools and ability, I'd start tackling door latches next. This tool does just that. From the "push" side of a door (turn the handle and push to open) the latch on the door is visibly and physically protected by that spine that runs the length of the door frame. A flexible tool like this can "swim" or "snake" its way into the doorjamb around the door and the frame and it can (often) leverage a latch into the "open" position. In a pinch, an old hotel room key card can perform the same function but lacks some of the length and "reach" of this tool setup. Don't use a credit card for this technique as it may damage the magnetic strip and the chips inside. (Example of this tool in action shown previously on page 21.)

- **Multipass**. $8. Available at UTAC's web store as well as many other fine retailers (I get mine from LockPick-Tools.com). Another great latch bypass tool. This one works from the "pull" side of the door. Use the "beak" of the tool to slip over the latch within the gap between the door and the frame, and rock the latch into the receded "open" position to access a single latch door. These can be found from multiple manufacturers at different price points and different make/model designs (some more effective than others).

- **Covert Jigglers**. $9. Available at LockPickTools.com and other retailers. These help tackle wafer-type locks. Arguably the next most common lock type mechanism in the U.S. (besides "combination" locks). You can greatly increase your opening capability with these tools with dedicated practice and research but, they are incredibly simple to use for low-security wafer locks, just insert and twist and jiggle. These types of "wafer jigglers" come in many types of makes/models at many price points and many different designs.

- **4-Pin Comb**. $8. Available at LockPickTools.com and other retailers. This comes with needing to do a little research on your target lock, as well as understanding what this tool is actually doing inside the lock. Once you understand this concept and have completed multiple entries, this tool becomes naturally easy if you pair your correct make/model of comb to the correct make/model of keyway. But there is a bit of a learning curve with this tool. Like most lock picking tools, the marketing can be misleading. This tool is NOT simply "plug and play". You have to "learn" how to use it.

- **EZ Decoders**. $10 for a two-pack. Available at UTAC's web store as well as many other retailers. Be careful(!) when ordering these online as they look almost identical to a product called "Quick Sticks". The two products have the same profile shape but very different thicknesses and are not interchangeable. I have found Quick Sticks have a low percentage of usability for which locks they'll work on, while EZ Decoders have proven themselves to work on *so many* different combination locks. Additionally, the fact that EZ Decoders become damaged easily, means that I stock my First Line Gear with only two EZ Decoders. Some retailers sell one of each (one Quick Stick and one EZ Decoder) in a pack, I'd rather toss the quick stick out and carry two EZ Decoders.

- **Hair Clip**. $1. Multiple sources. This tool can be used for restraint escapes and for some lock opening techniques. Not necessarily a required piece of gear, but it still has a home in the UTAC First Line Gear kit because it goes well with the blocks of instruction that we teach in our in-person courses.

Taping tools to the back of credit cards is a time honored method that is not to be scoffed at. Wallets and purses were designed to hold credit cards, not lock picking tools. Tape keeps your tools where your cards are without having them fall out. Also, where your cards go, your tools go. Using this method, you don't have to say "I'll grab my keys, my wallet, my phone, my gun, AND my lock picking kit," you can just say "I'll grab my keys, my wallet, my phone, and my gun (and my tools are always in my wallet)." This way, the "novelty" doesn't wear off of having to ALSO grab your lock picking tools every time you leave the house.

Scotch tape rips easily and doesn't retain great hold over time. Duct tape is way too sticky and leaves a residue and doesn't play well with flatness. Baby bear's porridge was too cold, Papa bear's porridge was too hot, go find tools and carry methods that are the Momma bear - for you. "Juuuuust right."

TACTICAL LOCK PICKING is not about getting you through *every* obstacle. That is economically an impossible feat. Again, good luck carrying every single lock picking tool known to civilization. TLP is about getting you through the highest percentage of reasonably openable locked obstacles, and being *prepared* to do so during an emergency.

[Above: an EZ Decoder (very thin flexible piece of metal "feeler" gauge) being used to decode a common combination lock at a UTAC Covert Entry Operations course. Opposite: same.]

SECOND LINE GEAR

Second Line Gear is defined as "A known packing list of tools that are stored in a small "go-bag".

A known list of tools is important because you will want the ability to immediately know if you have the necessary tool with you when looking at a locked obstacle or specific attack vector. Having a bag with a random selection of tools, or a selection of tools that you don't practice with or train will present problems in the field. You'll want to be able to quickly glance at an attack vector and to be able to immediately say out loud "I have the tool for that in my bag" instead of "I wonder if I have something for that." This is the difference between who is a master at their craft, and someone that is applying a hobby during an emergency.

A small go bag is important here, but only to start, and only as a guideline. The goal is not to shove as many tools into as large of a bag as possible. The goal is to know which tools are in your bag, know how to use them, and be able to deploy (verb) your tools into the field. A smaller shoulder bag or sling pouch helps you stay mobile over something like an old school toolbox or something like a fishing tackle box. Imagine climbing a vertical two-story ladder with a fishing tackle box in one hand, vs climbing with a messenger bag slung across your body. Big difference.

This is an unusual case in the tactical training world at large but you get to design this skill set, you get to select the gear that you purchase, you get to select which specific skills you buy tools for.

I recommend resisting the urge to change your loadout every week or even every month, or even after every operation. Your Second Line Gear loadout should certainly change and evolve but if you switch it around too frequently then you're risking introducing a new requirement to "search" through your bag to see if you actually still have whichever required tool in the heat of an emergency.

In your Second Line Gear loadout, is where you'll start to be able to expand in a few very unique ways.

You'll notice that you start adding tools that are specifically lock entry tools, and tools that are logistical/administrative/mechanical in nature, like screwdrivers and wedges, etc. You'll notice that of those tools you have professional commercial grade tools, and tools that have other purposes but that act as lock entry tools (a wire coat hanger is not a commercial lock picking tool, but one lives in my vehicle specifically for lock opening tasks).

You'll also notice that you have tools that are different from other TLP practitioners due to your Area of Operations (AO). Someone who works as a vehicle repossessor will rightfully have a large number of vehicle access tools in their Second Line Bag, and someone who works in a very corporate environment may have a lot of digital access tools etc.

You'll also get to start shaping tool selections that help you shore up weak areas in your skill set, and tools that help you double down on your strengths. Example: I am a terrible Single Pin Picker so I carry a few pin-tumbler Lishi tools to help me if my best option is to pick the pins inside a residential keyway. This helps me shore up a weakness in my skill set. Conversely, I'm really good with math and codes and by carrying our UTAC Brute Force Notebook in my bag I can make even better use of my skills on scene when it comes to doing brute force code hacking.

There is no perfect balance out there, just go start exploring!

Note: Of the logistical tools (screwwdrivers, hammers, etc.) we will not notate makes/models prices or sources for purchase to save precious space in these pages.

Front Flat Zipper: (Misc. Admin)

- Multi-tool. Headlamp. Small Bolt Cutters. Etc.

Front Bottom Full Pouch: (& Misc. Admin)

- **Wafer Breaker Kit** with extra American Bypass tools. $67. Available from multiple Locksmith retailers. Bypass drivers are great, when they work. It has been known in the entry community that the engineering of this tool is not a perfect science and often tools will arrive right out of the box and not be able to work properly, or at all. But when they do work, they're great! Wafer breakers are used when an American padlock has a security wafer that prevents the bypass driver from reaching up into the lock internals. Break the wafer, then continue your bypass driver use.

- **Long Handle Comb Picks**. $39. LockPickTools.com. You need to do research to determine which comb tool to use on which padlock. One of the many things about Tactical Lock Picking that only comes with research, and practice, and time.

- **Traveler's Hook**. $11. UTAC's web store. If you are close to getting a latch to slip with a jim tool but need just a bit more leverage, Traveler's hook to the rescue!

- **Jiggler Expansion Set**. $35. Locksmith retailers. When you are up against a stubborn wafer lock and your First Line Gear jiggler tools just aren't enough.

- **Auto Tension Tool and Double-Sided Wafer Kit**. $30 (approx). Locksmith retailers. A really handy tool for vehicle door keyways, but less and less useful by percentage of cars on the road, due to more vehicle manufacturers switching away from wafer locks and to other types of locking mechanisms.

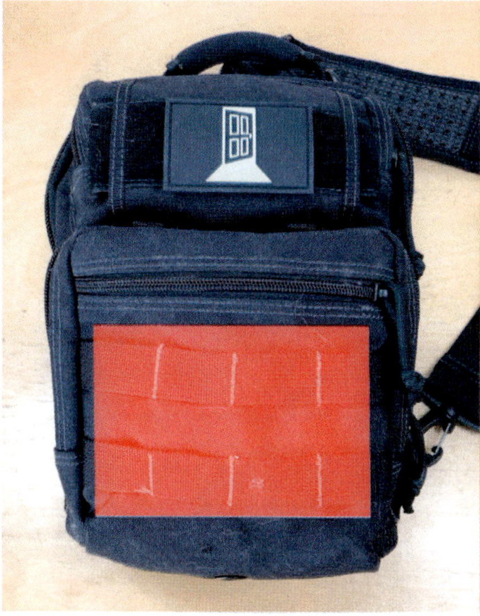

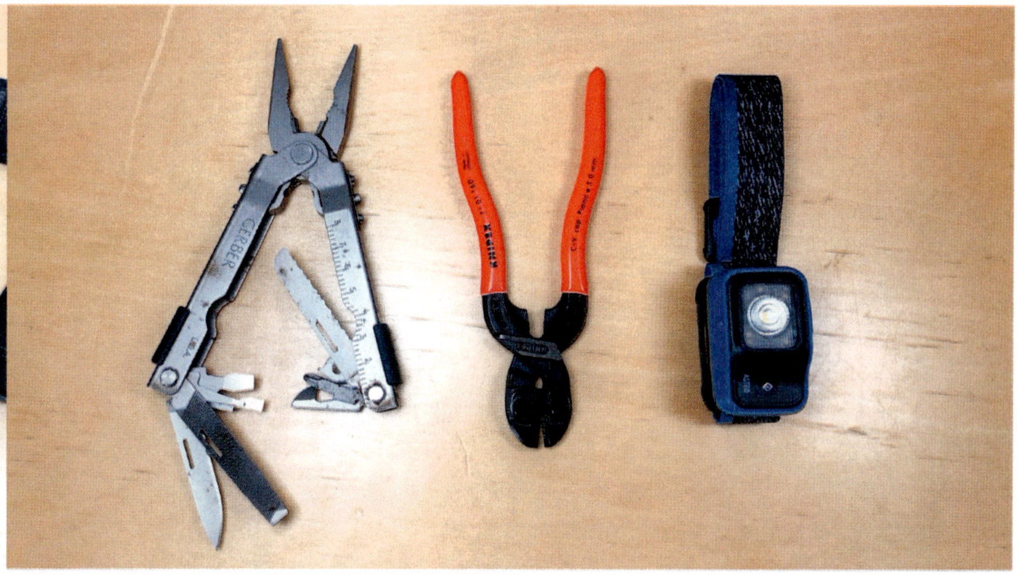

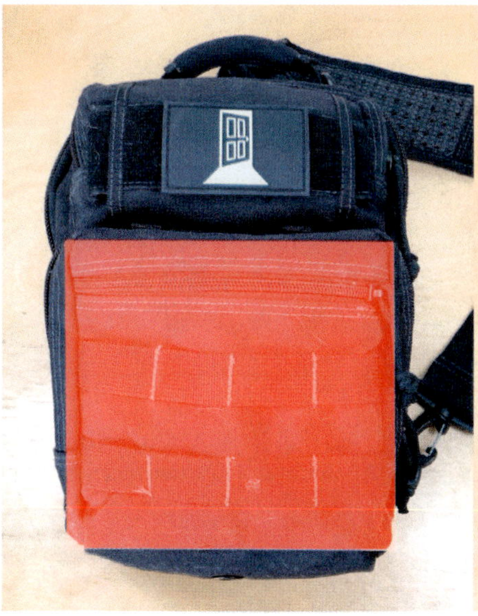

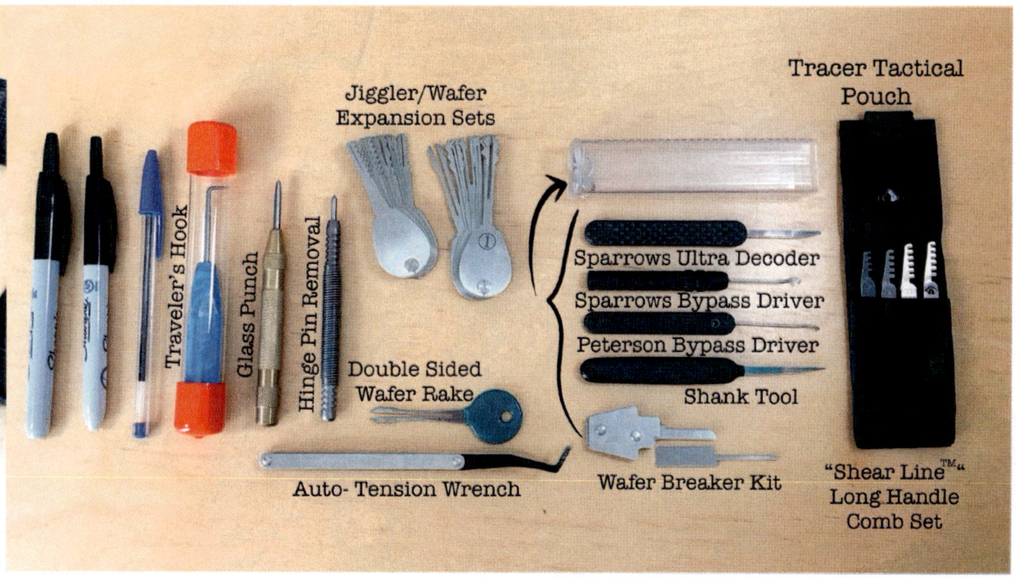

- **Glass Breaker**. Generic. Be careful with glass entries! Serious injury could occur to you or the people you're trying to save when you introduce sharp glass edges!

- **Hinge Pin Removal Tool**. Multiple makes/models retailers.

- **Sparrows Ultra Decoder**. $15. SparrowsLockPicks.com. In my opinion, Sparrows makes the best decoder tool in the game, Peterson is a close 2nd place with their set of three decoders in their bright orange handles. The EZ Decoders in our First Line Gear kit can accomplish a lot, tackling about 80% of openable or decodable locks, and the long handle gives a lot more tactile feel and dexterity to the user.

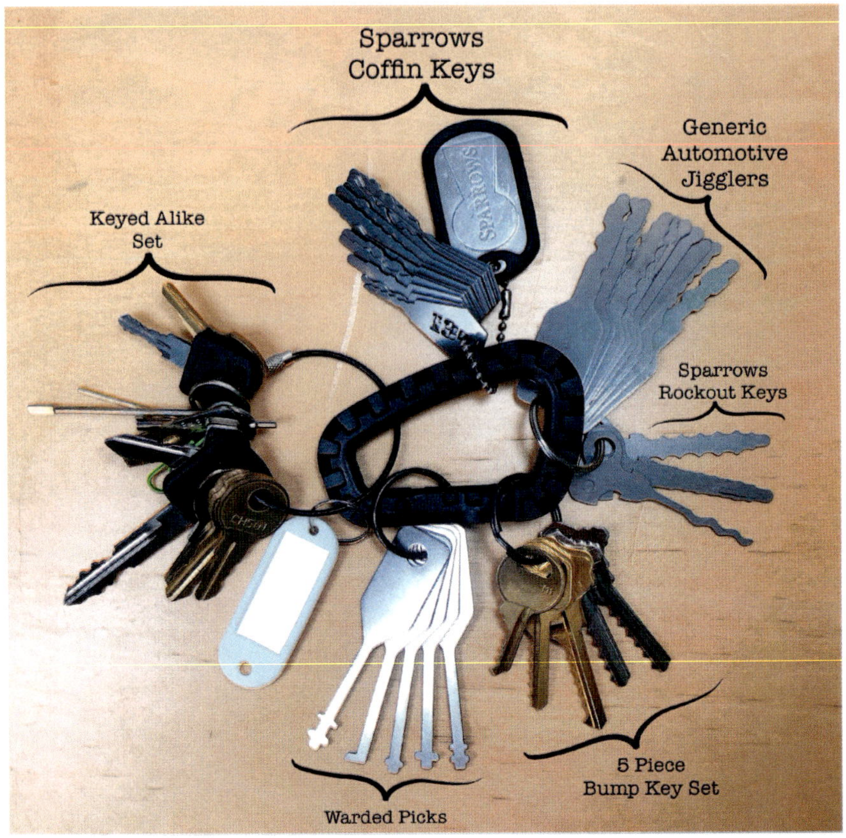

<u>Top Front Pouch</u>:

Everything below is stored on one plastic carabiner. This makes it so that I don't have to scramble around in my bag searching for the right key ring with the right size and shape of tools on it. I can just grab the whole carabiner and pull it out, hold it in front of my face, and remove the exact key ringed set that I need. Easy retrieval, easy stowage, organized operations.

- **Sparrows "Coffin Keys" set**. $27. SparrowsLockPicks.com. An awesome selection of full size and mini wafer jigglers. My only complaint is that the "coffin shape" at the bow of the tool kind of cuts into my fingers if I have to spend a long time jiggling the lock open, but the small profiles that the set provides is really worth the investment.

- **Sparrows "Rockout Keys"**. $12. SparrowsLockPicks.com. A great set to slide into an EDC setup, your First Line Gear setup. Not expensive, and they do their job. Doesn't hurt to throw them in your Second Line .

- **Generic Auto Jigglers**. $17. LockPickTools.com. The bow of the jigglers is very comfortable in your hands and it provides a large grippable area to hold on to. Great for stubborn wafer locks but doesn't allow for entry into small wafer keyways.

- **Warded Picks**. $13. UTAC's web store. Not the most common lock system type, by far. But, the set is inexpensive, easy to use, not very bulky, and when you do see a warded lock in the wild this set has a very high percentage of liklihood that the lock is able to be opened by this type of warded pick.

- **Keyed Alike Set**. Multiple prices and retailers. Or, source your own with a little research on some public message boards, the lists of keys and what they will commonly open is not hard to find. Depending on your AO, this set could be frequently used and immensely useful, or only useful in that small chance rare scenario.

- **Bump Keys**. $25. Some people have great success with this technique, I don't. One great usage of this tool is that if you are struggling to pick a lock, bump keys might give you a different type of edge for a possible entry.

Interior Zipper Pouch:

Same system as the carabiner, no loose tools, everythnig in a pouch that can be grabbed and pulled out, and held in front of my face on an entry so taht I don't have to dig through items at the bottom of a dark pocket.

- One pouch has **additional full sized lock picking "sets"** to give me a selection of different feeling rakes and hook picks, pocket sized and full sized. This pouch also has a large selection of advanced options for tension wrenches including a tension wrench style called "Top Of Keyway" (TOK) tension. This gives users a lot more room inside of a keyway to maneuver their pick around.

- Another pouch has **small o-rings** that help reset your bump keys faster instead of having to strike your bump key and then manually pull the bump key out one notch. The o-ring does that for you automatically.

- A last pouch has the **tips for a plug spinner tool**. I avoided adding a plug spinner to my kit for a long time but as other items have proven their track record to be less useful than expected, a plug spinner has proven itself to be more useful than I previously expected.

Note: Again, my kit changes about once or twice a year at maximum. You should always be reassessing your kit when big changes occur in some of these areas of your practice:

- Increased or decreased budget.
- Change in Area of Operations.
- New Technology for Entry Tools arrives or dissappears from the scene.
- A particular skill of yours increases or decreases and your gear changes to suit.

MINI STARTER
AirJack
1-800-323-TECH
MADE IN USA
KEY BLADE
SEARAT
Vehicle
Air Shim
Plug Spinner
Lishi Decoders
Door Latch Jim Tool
Bump Hammer
Vehicle Door Gap Tool

Main Storage Pocket:

- **Pin Tumbler Lishi Tools**. $80 per (approx). Multiple retailers. This is one of the biggest changes to the lock entry world in the last several years, hands down. It has taken the hard earned artform of Single Pin Picking and has made it much more "plug and play". Not entirely, but much more entry level of a skill. I carry two of the most common residential keyway makes/models, Kwikset (5-pin) and Schlage (5-pin). I also carry a Lishi tool for American 1100 padlocks, a very common padlock and quite a task to pick open for a beginner. I have never claimed to be the best lock picker in the world, I'm not even the best lock picker in this room and currently it's only me and my service dog "Arrow" here. I unabashedly use Lishi tools to give myself a bump up in capability for when I absolutely must use a pin-manipulating attack within a keyway.

- **Bump Hammer**. $16. UTAC's web store. Goes with our bump keys. Too hard of a hammer and too hard of a strike and you risk damaging the lock that you're attempting entry into. Too soft of a material and you won't get that rapping motion that the pins require.

- **Plug Spinner**. Multiple price points for multiple makes/models from multiple retailers. In theory, if you can pick a lock open to the left, you should be able to reset it and then pick it to the right. In theory. But, in theory, theory and practice are the same. In practice, they are not. A quote often attributed to many different historical figures. In the field, the field is not your couch. So when I accidentally unlock a keyway in the wrong direction and it keeps the lock "locked", or if I intentionally pick a lock in the wrong direction because I'm struggling to pick it in the right direction, a plug spinner quickly takes the keyway (the plug) from one side of the setting of the pins, and snaps the plug past the point where the pins are so fast that the pins don't have a chance to fall into the plug. If you are lost in this concept it is something that I promise you will quickly become familiar with when you start picking deadbolts and knobs open through the keyway.

- **SEARAT**. $100. Ignitionusa.us. I do not recommend this tool unless you are professionally operating with your Tactical Lock Picking skill set. Not as a statement of "gatekeeping" but because it is expensive, bulky, heavy, has a very sharp edge, and its function is able to be replicated with other makeshift items very easily. One of its most common purposes is to reach back behind latch proctector plates in order to slip a latch from the pull side of a door. It is a good tool, but I would weigh your purchase of it carefully.

- **Sidekick Door Gap Tool**. $30. Locksmith retailers. This tool is non-scratch (less-scratch) plastic and rubber and helps make vehicle entries easier by providing an initial creating of a gap between a vehicle's door and the frame/body of the vehicle. This helps make it much easier to get your airbag in place, and much faster.

- **Mini Air Jack**. $40. Locksmith retailers. You can find variations of these at most local hardware stores as "air shims" for around $20 each. The one I carry in my bag has a woven easy-slide fabric on it and the ones at hardware stores are often rubber coated. The fabric ones slide into vehicle door/frame gaps much better than the rubber air shims. I also keep additional air shims in my Third Line Gear, but it's nice to at least have one with me in my Second Line Gear.

- **Sparrows Quick Jim**. $11. SparrowsLockPicks.com. This, and the generic versions of this tool, are all pretty similar in capability. I really like the handle design and material on this Sparrows version. Used for slipping a door latch from the pull-side of a door.

A word on titles: We all know someone who has a qualification hanging on their wall who is as dumb as a bag of hammers. And we all know someone who is arguably one of the best in the world at their "thing" but either they don't have the paperwork qualification for it or maybe a qualification doesn't even exist. If you agree with me that that much is true then maybe we can agree that qualifications are tools for bureaucracies and capability is the only measure of competence for those who choose to be the masters of their craft. Thank you for coming to my talk.

Back Pocket:

- **DoorJamm**. $12. UTAC's web store. One of my favorite tools in the "Low cost, high value, applicable to multiple make/model obstacles" category. You can also order custom artwork straight from the DoorJamm.com website. These perform multiple functions. They keep an auto-locking door from automatically locking after you gain access (think hotel room doors and commercial perimeter doors etc.) and they also let your backup/team/etc. know which door you have accessed and went into, AND it allows those support members to not have to gain their own access on their way in to assist you.

- **Commercial Door Latch Shims**. Multiple prices and retailers. These are fantastic tools that no Second Line Gear bag should be without

- **Brute Force Notebook**. $30. **Only** available at UTAC's web store. This notebook lists every single possible code to a number of mechanical push button locks. There is a lot to this product but a short explanation here is that because on many mechanical push button locks, if "1-2-3-4" opens the lock, often inputting "4-3-2-1" will also open the same lock as a proper code. We have boiled down the useless entry codes for many locks resulting in a list of the smallest number of combos you'd have to try to open a few different makes/models of push button combo locks. TLP is a numbers game and a percentage game. A brute force attack (trying every possible code) could be a very fast or a very slow technique but this is all weighed against your mission variables. 15 minutes to try every possible code might be an eternity on some operations, or it might be one of the fastest possible vectors of attack depending on your variables and your environment.

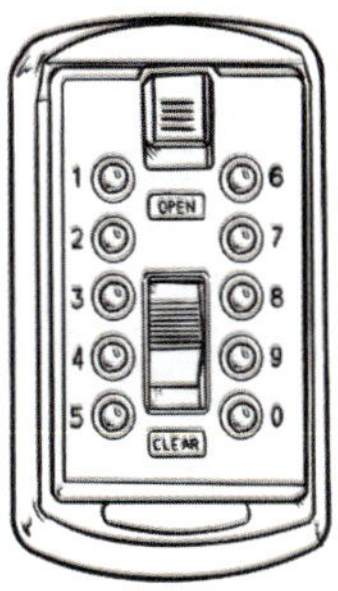

EVERY

POSSIBLE

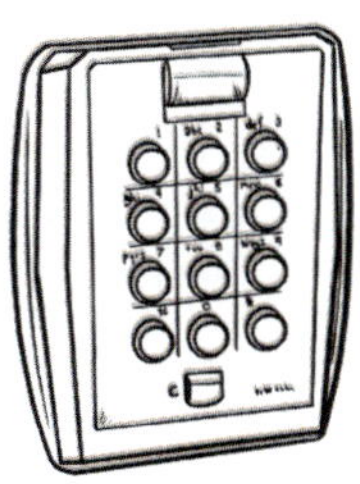

CODE

*That is kind of a big deal...

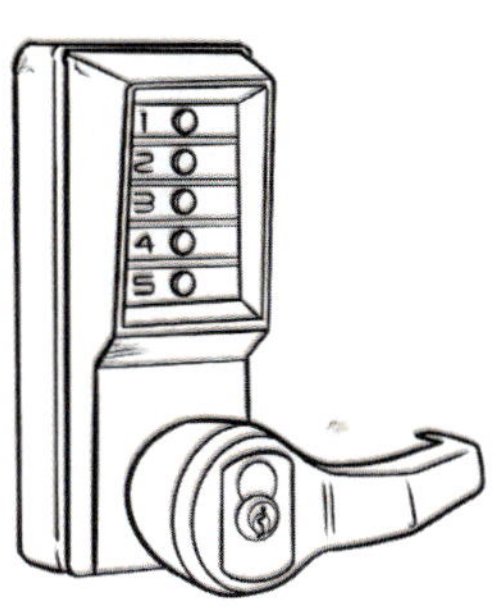

THIRD LINE GEAR

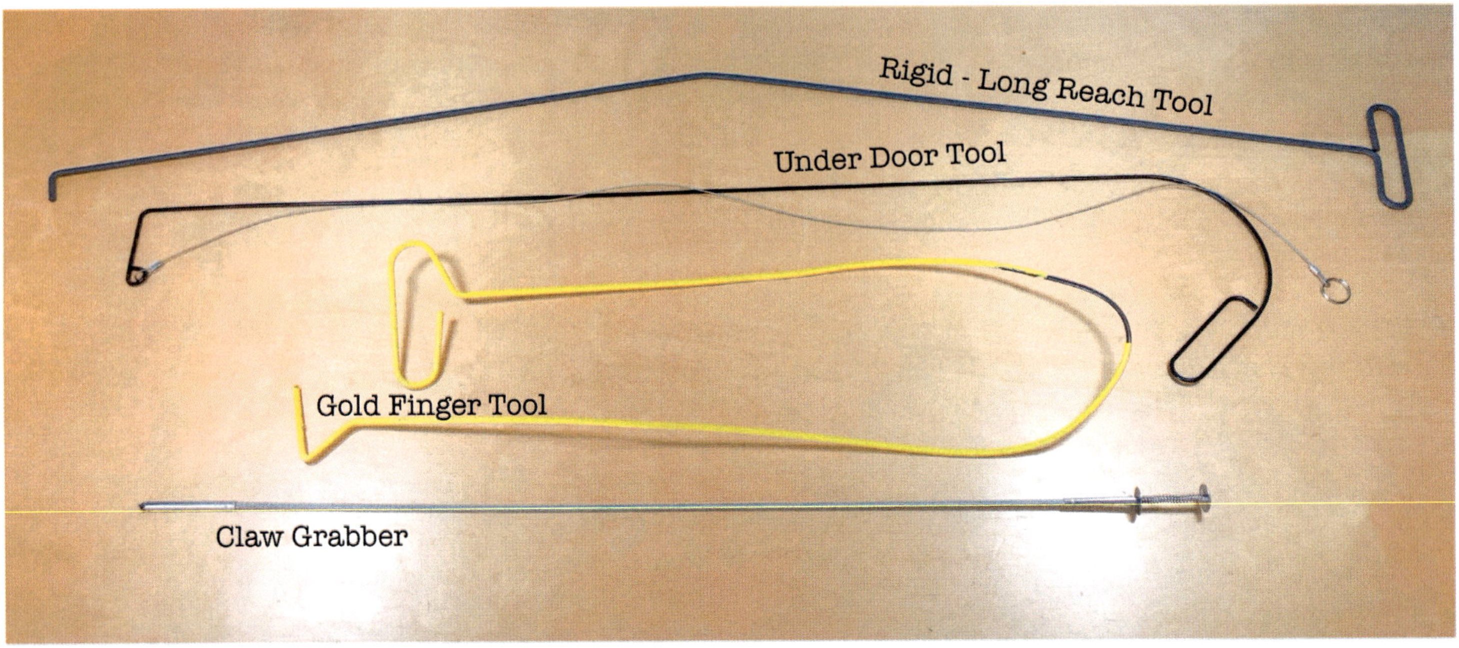

Third Line Gear is defined as "Tools that are large, bulky, specialty, and otherwise don't have a place in your first two lines of gear".

Where these tools live are commonly in two places, in the trunk of a vehicle or in a closet/garage (home or at work). Other creative options are allowed and encouraged as long as the storage system works for you. Remember that these Lines of Gear, and all of our curriculum, are designed specifically for you to alter to suit your needs.

What we commonly carry in our **Third Line of Gear**:

- **The "Gold Finger" Reach Tool**. $27. Locksmith retailers. This is a long thick gauge wire, coated in yellow, and it is semi-rigid. The pro is that you can bend it for easy storage and you can also custom bend it during a vehicle entry to reach a hard to reach spot within a vehicle. The con is that because it is semi-rigid it is lacking in the area of "pushing" buttons. The tool often bends and slipps off of buttons that require a lot of pushing force.

- **Rigid Reach Tool**. Multiple prices for makes/models from different retailers. As a more rigid tool, these can not be bent for storage and can not take custom bends during entry to reach awkward angles. But, when a button needs to be "pushed" the rigidness of this tool far outweighs the ability of the flexible "Gold Finger".

- **The Under Door Tool**. Multiple prices from multiple retailers. One of the best tools in the game. While this tool is "simple" in its application process, it is not always "easy" to gain immediate entry. Social media may show you that this tool always works, and always on the first try, it can sometimes present a very frustrating challenge to work through some of the troubleshooting required. Still, this tool is 10/10 recommended from us.

- **Claw Grabber**. Multiple prices from multiple retailers. Remember the game that you played as a kid where you put a quarter in the machine and then you controlled the big sky-claw in attempts to grab stuffed animals from the pit? Like that, but for vehilce entries to pull buttons. Not the tool that sees the most use from me, but sometimes it's a great option.

- **Sledge hammers and pry tools etc.** Multiple prices and retailers. Yes, this book is dediated to "Tactical Lock Picking" but we don't exist in a vacuum. If we are practicing and preparing for *emergency* entry in real life, we can *specialize* in non-destructive tools/techniques but we should still be prepared to make entry in a more kinetic fashion when the need arises.

[Pictured: ...and a ram is not just a ram the same as a picking tool is not just a pick. Makes and models have pros and cons, and nuance still reigns supreme.]

FOURTH LINE GEAR

Fourth Line Gear is defined simply as "Information".

We've been using this concept for a few years now, and we're happy that we get to add this as a new concept to this 2nd Edition of our foundational Tactical Lock Picking book.

Let's say you had something like a digital PDF file that lists which locks can be opened by which comb tools. You might be able to access that via your smartphone, and your smartphone might be a part of your EDC, sure. What about UTAC's Brute Force Notebook featured in our Second Line Gear? It's a book! Certainly a tool that is used because of the information it contains. What about a saved playlist of the dozens and dozens, or even hundreds of videos online of people using unique tools and techniques to open a wealth of different locked obstacles, that would certainly be a resource that you'd be able to use on scene in order to help you make an entry on a lock that you remember seeing in some video somewhere that you saved to a playlist.

Don't get stuck on the "where" for this Line of Gear. It could be photos or videos saved on your phone or on an internet playlist. It might be a lock picking message board that could provide you an answer at 2am. It could also be a "phone a friend" that you call when you have a unique problem in front of you. It could be something like our Brute Force Journal that is stored in your Second Line Gear bag but it could also be your growing library of lock picking books at home or on your office bookshelf.

Start thinking about adding "information resources" into your Lines of Gear, wherever they happen to be physically, or digitally, or even mentally stored!

Examples:

- Documents listing which tool will open which makes and models of locks.
- Video playlists of lock picking techniques.
- Notebooks listing out brute force codes.
- Useful web pages stored in a bookmarks folder.
- Having a memorized Target Assessment steps guideline.
- Learning and then knowing how to use math to determine how many combinations a given combination lock has.

DISPOSABLE TOOLS

Allow us to give you a new definition for "disposable". In our practice, it does not mean "low quality" or "unimportant".

We use "disposable" as a term to describe a type of tool that may fit just one of the elements of our definition below, or many of the elements.

- Tools that are low cost. (Although some high-cost tools can still be considered "disposable" in our craft.)
- Easily damaged - and therefore requiring frequent replacement.
- A mindset for a tool that when it is damaged I don't feel bad about it because it fits this rough guideline.
- Tools that are so uniquely valuable that it makes sense to *buy* and own two of them.
- A tool that makes sense to *carry* multiples in the field.
- A tool that makes sense to always buy two if you lose or damage one (building a stock of them up).

EZ Decoders are a tool that fits every definition above. They are low cost, they are easily damaged, it makes sense to buy and own two (or more) of them, it makes sense to have more than one that you carry into the field (so not just having a backup at home but *carrying* multiples), and it makes sense to have a stockpile of them that is constantly growing.

The original Bogota's, the classic two piece set. At $25 they are not "expensive" standalone, but when you have to buy multiples it starts to add up. For me, they are pretty low cost. They don't necessarily damage easily but overtime they start to bend and warp a little with heavy use. They are in fact such a high capability item that it makes sense to have an extra set (at least a set sitting at home). I don't necessarily carry two sets in the field. And I don't always buy two of them every time when I damage one of them, but I do have a small reserve of them at home.

Commercial Door Latch Shims. Low cost. Somewhat easily damaged (they stop looking "like new" with their first use usually). I do not feel bad when one of them shows major signs of damage. They are very easy to carry multiples in the field and it makes sense to have at least two of them in the field of varying thicknesses. These usually come in a set of six (three thin and three thick gauge) so I don't build up a reserve of these, but when I get down to about two shims left I'll buy another pack of six.

Lishi Decoders. Expensive, at around $80 per decoder. The tips do break with heavy usage. I don't carry two of the same model in the field but I do have a requirement to have at least one backup at home for immediate replacement day-of if one breaks during an entry. I do not have a growing stockpile of them with each additional replacement order. Uniquely though, I do not feel bad when one breaks in the field. They are so incredibly valuable to me that as soon as one breaks I immediately log into one of my web stores and I don't waste any time ordering an immediate replacement. I mentally treat that loss as a disposable and it gets immediate replacement.

Hopefully this gives you some ideas for managing your inventory, maintenance and replacement procedures to help give you a depth of craft.

Modular Options

Smartkey Field Tryout Set: $445. **Only** available at UTAC's web store.

Designed by the UTAC team, from scratch, a custom designed fanny pack that holds 243 keys used to access Kwikset Smartkey locks. Not something that you'd wear around your waist every day. Probably not even something that lives inside of your Second Line Gear bag. But worth keeping in a vehicle trunk or in your gear closet, so that you can grab or have it delivered to you when you make the determination that you will be attempting entry through a Kwikset Smartkey lock.

Remember that urgency timelines are relative. This tool set takes around a maximum of 20 minutes to try every single key in a Smartkey keyway. It is possible that the last key that you try could unlock your target lock but it is also possible that the first key that you try could unlock your target lock. It's a 1/243 chance.

Even for something as simple as "try every key" there is still a boat load of nuance of craft to be considered.

We pride ourselves on teaching "Software over Hardware" or valuing training, knowledge, and experience, over having the perfect selection of tools.

[Photo below: Our "Kwikset Smartkey Field Tryout Set" which is a modular grab-and-go set that you would grab from your car or your storage location if you know ahead of time that you'll need it, or after you discover that it is one of your best reasonable entry methods. Available in our online gear store now.]

Key Cutting Fanny Pack:

Small modular bags are a great start to storing slightly larger and more varied tools. They're also a great way to store themes of tools like the KeyGen tool selection pictured above, or the tryout key set on the opposite page.

While the above is not a tool for lock picking but a selection of tools used for key generation, using small modular pouches is a great way to organize your loadout for entries. (Our book *Tactical Lock Picking Vol. 2 - Field Key Generation* teaches this key cutting skill set in depth.)

CONCLUSION

If I had to give you this conclusion in twenty words or less I'd tell you to remember "Software over Hardware" and I'd tell you that "Gear storage and deployment methods are a guideline, not a rule."

Gear is fun. Gear is cool. And yes, gear is very important but, don't get wrapped around the axle trying to spend yourself into debt just in case the zombies come.

Think of this skill set more like a martial art.

> You show up, you go to training, and maybe you spend some time learning knife fighting concepts. As you're progressing, you'll start to religiously carry a reliable knife in your pocket as an EDC. Maybe you also start shooting at the range frequently with a good instructor and you start to EDC a concealed handgun everywhere you go. Ask yourself: Gun - Knife - Phone - Wallet - Car Keys - AND knowing how to use them all better than the average person, is that a pretty bad ass setup? In my humble opinion, that can solve a LOT of problems.
>
> Now imagine, conversely, someone that goes to your same martial arts studio who always has a new training bag stuffed with flashy knife trainers and with colorful wrapped bamboo sticks, and boldly colored boxing mitts, etc. etc. but he skipps class a lot and doesn't really listen when the instructor talks. All that flashy gear isn't going to help him. Imagine also the guy at the range that has the $10,000 rifle with the $10,000 scope on top, bipod, lasers, top of the line flashlights attached, all color coordinated, but can't consistently hit a target at 100 yards, while you've spent years honing your craft from every conceivable angle with your factory model $1,000 rifle with its $2,000 red dot sight and a $600 flashlight - and thousands of hours and thousands of dollars worth of training. Who is going to be a better problem solver in a gunfight?

Remember too, that this curriculum is not an exclusive curriculum. There is nothing in our teaching that prohibits you from thowing away almost all of the recommendations and guidelines that we provide you, and adding in your own useful tools/techniques/tactics etc.

Our goal at UTAC, arguably our biggest goal, is that we want to help people to start to think differently. Even two operators on the same response team should have different TLP loadouts. They should look like overlapping venn diagrams but they should also celebrate their outside edges. Example:

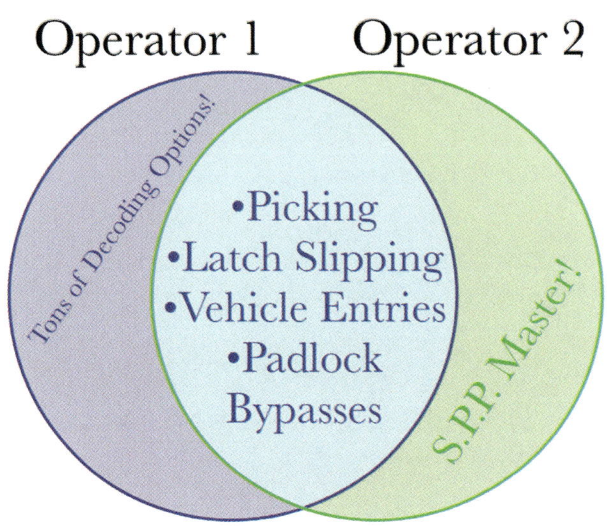

Your new gear loadout will always feel super cool when you initially design it... until you bring that new loadout design on your first operation!

Go train. Go fail. Go learn.

CASE STUDY: SUICIDE BY PILLS

"Three-William-Sixty-One," can you be 51 to a Signal 32? This was called in by the adult daughter of the victim. Victim is locked in the master bedroom with a note outside the door. Daughter can't get into the room to assess."

"Dispatch, I'll be 51. Priority 2," I responded and started driving quickly with my lights and sirens on, since this was called in as a suicide attempt. Time was a factor.

STEP 1 & 2: MRE & LOU

I would have to confirm and adjust on scene, but it would appear that we already had MRE because (1) a family member on scene would be allowing us entry and (2) a note on a door indicated a likely suicide attempt, so we'd be going in to save a life. It would also appear that we'd have a degree of heightened levels of urgency to determine if the victim was still alive in their locked bedroom.

I arrived on scene at the same time that Deputy J pulled up. I grabbed my Second Line Gear bag, and the two of us walked briskly toward the first-floor apartment. If the few small tools I had in my First Line Gear in my pocket weren't enough to properly attack this locked problem, it would be nice to avoid having to go back and forth to my vehicle for more gear, wasting precious time.

The 911 caller, the victim's daughter, was standing outside her mother's apartment front door. She told us, "My mom lives by herself in this apartment, and we usually chat once a week. She missed our phone call today and wouldn't answer her phone. I showed up with my keys and entered her apartment, but she was locked in the master bedroom with a note hanging on the door." Roughly, the note read "[I took a bunch of pills, don't try to stop me. I'm probably dead already.]"

With the daughter's permission and a reasonable expectation this was a legitimate suicide attempt, we entered the apartment to do what we could to help.

I checked the corners and closets and peeked into empty rooms as I made my way toward the master bedroom just to ensure that we'd have a safe scene while we tried to manage our access to the locked master bedroom. As I stood at the master bedroom door I observed that the door, which was a push door with a privacy knob, was locked and the note was hanging on it. Adjacent to my left, a door opened into a hallway bathroom.

[Photo: opposite]

STEP 3: ATTACK VECTOR IDENTIFICATION

At the time, I thought the only option for entry was through the master bedroom door. It was a privacy door - a basic, plain "vanilla" door - top to bottom. It was also a push door that could be kicked in with minimal effort. It had one doorknob, a privacy knob, on the left-hand side. The knob's only feature was a small hole in the center. I knew I could attempt a shim technique to slip the latch from the push side of the door. I also knew that, at the time, I did not have a tiny screwdriver or a privacy door "emergency key" in my kit (a tiny brass tool shaped like a poker or shaped like a tiny flathead screwdriver).

Privacy Knob defined. A doorknob with a simple push-button lock or a twist lock on the inside and no keyway on the outside. Sometimes with a small hole on the outside (or other device) for emergency unlock access.

My partner on the call, Deputy J, was now standing shoulder to shoulder with me and one of our sergeants was stationed behind us. Fire and EMS had just arrived on scene. One of the responders had a heavy axe and a Halligan, one in each hand, and another was carrying a medical response bag.

[All photos in this case study are **recreations**.]

STEP 4: MANUAL UNLOCK CHECK (X2)

I banged on the bedroom door and as loudly and politely as possible delivered the boilerplate "Hello! Are you in there? Are you okay? Sheriff's Office. We're here to help you with some nice EMS folks! Can you hear me? Can you respond to me?"

I took the first physical step for my on-scene entry: I manually checked to make sure the obstacle wasn't already unlocked. Then I took a deep breath. Then I tried again, even harder (up, down, left, right, clockwise, counter-clockwise, push, pull, etc.). It was in fact locked.

I was able to hear a TV or radio through the door, and I was also able to hear someone speaking though they were not very clear or coherent.

Right away Deputy J asked, "Hey, Sarge, you all right if I kick this door in? We'll have this call wrapped up in two minutes flat." The sergeant considered, then approved the move.

I ventured a suggestion. "Hey Sarge" I said, "I'm all for letting Big J here get his exercise, but let me run something by you first. This would not be the first time in history someone attempted suicide but then changed their mind and decided they now want help. And if this woman's sick to her stomach from pills and starts crawling toward the door to unlock it for us, or if she's lying there now with her head facing the door and we kick the door in, there's a good chance we might kick the door right into her head. How about this? If I can't open the door with my tools, then we let Bubba J here kick it in." The sergeant considered the additional info and then approved the tool entry attempt first.

[The recreation photo on the previous page shows a large gap under the door. During the real-world scenario the door was flush with the ground and looking under the door for visual information was not possible.]

STEPS 5 & 6: TOOL AVAILABILITY & NOW OR LATER

I had my Second Line Gear bag with me and this was a pretty simple low-security entry so now was the time. Right? What else could I possibly assess? (Famous last words.) I started a loose mental timer to track how long I was going to be spending on this locked obstacle.

The Target Assessment phase was now on hold while I started my physical entry attempt. I quickly pulled a clear plastic commercial shim out of the bag at my feet. At first it was a little tough getting the shim into the door jamb.

[Left: Sometimes, if it's too tight of a fit to slide your latch shim directly into the door frame near where the latch height is, you can usually push in the door at the top corner or the bottom corner and it will "give" a little bit and create some more room to be able to get your latch shim slid between the door and the frame. Then you can focus on sliding the latch to the correct height.]

Some very high security facilities make it very easy to use a shim or a jim attack on their door latches because of certain bureaucratic functions for facility design and maintenance procedures, while some very low security homes and apartments are very difficult to fit a shim or a jim tool into a tight tolerance between the door and the frame. The illusion of security is a two-sided game. Sometimes appearances are correct and sometimes they are quite deceiving.

I now had the shim right up against the latch. I applied both a pushing and a pulling force on the door while continuing to try to drive the shim through the latch. No luck. I started to feel a little embarrassed. I knew this was a cheap door. I knew the way into the room was through the simple privacy latch. I also knew the only other option was a tiny screwdriver or emergency key that at the time I didn't have on me or in my bag. I continued to struggle.

Meanwhile the sergeant meandered around the corner.

[Above: Within two steps of the locked obstacle (the master bedroom doorknob) were a few nooks and crannies that are slightly above toddler-height. The victim was a grandmother who sometimes babysat her granddaughter. Hiding things far back on a medium or high counter is a common occurrance.]

"How long is that gonna take there, hot shot?" My country bumpkin partner was happy to prod me.

"Dude, I know that I can do this. I've done this a hundred times before on higher security doors," I replied. "Shitty door" I muttered under my breath.

I felt a tap on my shoulder. I was not in the mood to hear whatever Deputy J had to add, but it turns out my door-kicking partner was not the one tapping. It was the sergeant. He extended his hand toward me, palm facing up, and on it was an emergency key. He wasn't mocking me; he wasn't all puffed up about how smart he was; he simply said, "Here ya go."

CONCLUSION

Three seconds later the door was unlocked, EMS went in to assess the victim, and I learned a few things about myself and about this skill set and how I could make even sharper systemized decisions in the future. I really like this case study because the flow was pretty straightforward. The Target Assessment phase proceeded almost perfectly in order, and I remembered to do quite a few things right, although certainly not everything.

I also really like this case study because of all the emotion and the human interaction between the caller and the victim, the responding officers and the people on scene, and even between the responding agencies and between me and my fellow Deputies.

I also really like highlighting the surprising fact that inexpensive infrastructure can sometimes have unintended security features, like a tight-fitting and sticky door jamb, which can hinder entry specialists. Conversely, I've used a laminated sheet of paper to perform this same shim technique to gain entry into a secure multimillion-dollar government headquarters building.

The door latch was not the only way into the bedroom. The apartment was on the first floor which meant that while I was working on the lock, I could have sent my partner around the outside to see if the window was able to be slid open, or at least if he could see the location and the consciousness level of the victim. Not utilizing my resources in this fashion was a failure of mine that now helps to strengthen our curriculum when we teach others.

I knew what tool could have opened that door but I didn't have one in my kit, although I did have several at home. Such a tiny and insignificant seeming tool could have made all the difference earlier on in the call. I now carry the privacy knob "emergency keys" in my kit for the two biggest residential door hardware companies in the U.S. (Schlage and Kwikset - and their knockoffs and cloned locks) and I also carry two small screwdrivers (flathead and Phillips) that can also perform the same function, in my second line gear bag.

I was *well* aware of the two-step rule at the time: "If there's a hidden key or a code written down nearby, it's usually within two steps of the locked obstacle." And even though the sergeant ended up looking for that key, it was still my failure for not specifically sending someone on that search as a part of selecting and directing during my entry. You could say that I got "sucked in" aka "tunnel vision" on the tool and technique I was applying but at least not completely. I did still have a mental timer going in the back of my mind, and there is still a good possibility that I could have gotten that latch sometime soon. Hard to say for sure what would have happened.

Makeshift tools are always something to consider as well. We had more than a handful of responders on scene, and we were inside someone's apartment. If I asked all of the responders, and the victim's daughter (familiar with the apartment) for a tiny screwdriver, one might have been found pretty quickly for an almost certain quick entry.

I also learned a little humility on this call. There's no telling whether I was just about to get that latch to slip or whether I never would have gotten it but, someone else did solve the problem before I did while I felt confident that I was the entry expert on scene. When you're right, you're right. And this time, the sergeant was more right-er than me, at least he was right more quickly than me.

I am still happy that we didn't kick the door in right away. I still call that a logistical win as far as managing the scene and keeping the mission in focus.

Everybody has good days and bad days. Everyone misses things sometimes. I am proud of the fact that failures shape a very large part of our curriculum that we teach. We live in a strange culture where failures are frowned upon, and I try to do my best to show students, readers, etc. the incredible value of failures.

CURRICULUM: TWO-STEP RULE

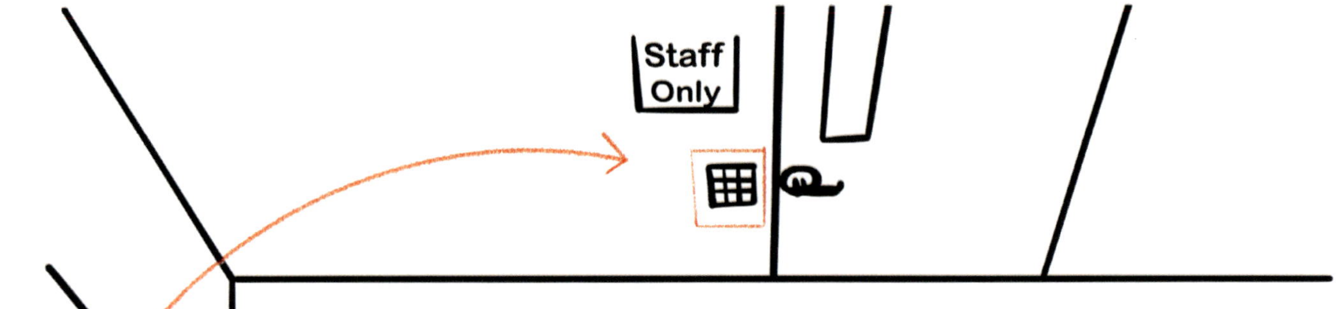

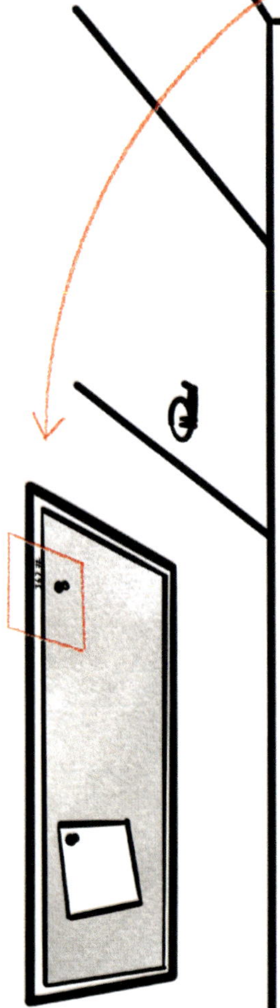

If there is a hidden key or a code written down for a locked obstacle, it is almost always within two steps of the locked obstacle.

It may make a lot of sense to spend your first moments on scene looking for a hidden key somewhere within two steps or so of your locked obstacle. Or it may not be that necessary until you start to troubleshoot your entry if you start to struggle.

Hidden keys are very common in certain environments and not so common in others. It would make sense to look for a hidden key first if the lock would require more than a 10-second entry and if the environment is a likely candidate for having a stored key.

Some environments that signal that a key is likely to be stored nearby:

Residential homes with a private or semiprivate front porch having lots of nooks and crannies where small items can be hidden.

Apartments with separate (not linked) entryways.

Office desk drawers (whether the drawer is locked or not). Most standard locked desk drawers are held shut with inexpensive and low-security wafer locks; in a pinch a few paperclips will open them.

Work areas accessed by many employees. A 24-hour facility that workers need to access at different times of the day or night is more likely to have stored or hidden keys.

If there are extra hands on scene or if the facility is large and a full perimeter check would take a while, I'd send someone to look for hidden keys and codes while the entry expert starts using their physical craft. I'd also check around an entry area for a minute or two after each unsuccessful attack on a specific vector and before moving on to the next.

You might not need to check the two-step area at all, depending on whether you're pressed for time or whether you're certain of making a successful entry. If I identify an easy entry point and I have the right tool in hand and if entry would take me fewer than 60 seconds, start to finish, it would be silly to spend five minutes looking high and low for a hidden key. Again, the economics of time management is a huge factor for us in Tactical Lock Picking.

[Right: "Staff Only" Lounge at a trauma center. If I had to gain access to provide emergency services through that locked door:
-Slip the latch.
-Pick the keyway.
-Under Door Tool entry.
-Go find someone with RFID Access, or who knows the code.
-Guess the code.
-Smash the window and reach through.
-*Go find a possible second entry door/window first!]

[Below: A cork board, announcement board, two steps to the left of the keypad shown in the photo on the right.]

[Right: Not only is the code to the door written on the edge of the cork board but, I could imagine the speech that all of the current staff gives to every single new hire employee "...and if you forget your RFID badge somewhere, just type in the code...and if you forget the code it's on the cork board. Just look for the clear push-pin in the top right corner." More on this same case study later.]

The Two-Step Rule is nothing new and nothing proprictary. I first came across it in a book at least a decade ago and I've tried to find the source to no avail. But it is another one of those items that once you hear it, it is registered in your brain as "I already knew that, but I didn't know that I already knew that."

If someone does store a key to their front door somewhere outside the home, the hiding place is rarely around the side of the house, over a locked gate, up a second story balcony, and under a 75-pound flowerpot. This is because security is always a balancing act between effectiveness and efficiency (a.k.a. The level of security and the ease of access. A.k.a. Security vs laziness). On returning home from a long, miserable day at work to discover you are locked out of your own front door, you probably wouldn't want to put too much effort into retrieving your spare key at the end of a physical obstacle course. You'd likely rather just reach under the doormat.

> **the three biggest indicators of whether a key is likely to be stored nearby a locked obstacle:**
> **(1) if the obstacle relatively hidden from the public view**
> **(2) if the environment is not clean and sterile, busy environments with less visibility and more nooks and crannies are ripe for hiding spots, and**
> **(3) if the obstacle requires shared access from different people at different times.**

Have you ever stored a key for something (or a written four-digit code or passcode) in a simple unlocked drawer and thought to yourself "This isn't the most secure place for a key, right up front in the front left corner of this drawer..." and then pushed the key (or code) to the back of the drawer for a little extra "security"? Maybe covered it with a scrap of paper? It's okay. This is not an attack on you. I've done the same thing hundreds of times! It's a calculated risk we all take, and the point is that the way people think about security often falls into certain patterns.

- Common examples of the Two-Step Rule:
- Key under the mat
- Key above the door frame or a railing
- Key under a flower pot
- Key in a top desk drawer
- Key in a mailbox
- Key inside a small box or enclosure near the door
- Key inside a front lawn/garden ornament
- Car key magnetically stored under the vehicle frame (this is less common due to key fobs granting access based on proximity)
- Code written on a white marker board or cork board near a keypad access door
- Codes written on sticky notes on or near a desk

And seriously. Listen, don't ask me why but:

I. Am. Telling. You.
When. You. See. A. Decorative. Frog. Near. A. Locked. Door.
Always. Always. Check for a key. Under that frog!

There has got to be some hidden informational code across all ages and all cultures in human DNA that says "Oh. This frog'll be a good spot to hide a key underneath."

The Ultimate Hitchhiker's Guide to the Galaxy

His house was still there.

How or why, he had no idea, but he had decided to go and have a look while he was waiting for the pub to empty so that he could go and ask the landlord for a bed for the night when everyone else had gone, and there it was.

He let himself in with a key he kept under a stone frog in the garden, hurriedly because, astoundingly, the phone was ringing.

He had heard it faintly all the way up the lane and had started to run as soon as he realized where the sound was coming from.

1830

So Long, and Thanks for All the Fish - Douglas Adams (1984)

[Below: And did anyone notice *this* little guy while reading the first chapter?]

KEY STORAGE BOXES

Key storage boxes definitely count within the Two-Step Rule guideline but in a unique way. They are not keys that are hidden nearby, they are keys that are intentionally visibly stored nearby.

Some residential and commercial structures have key storage boxes mounted on the outside of their structures, sometimes hanging on doorknobs in large padlock boxes or mounted on a wall or post or even slightly hidden sometimes. Sometimes they are plainly visible next to the front access door and sometimes they're around the side of a building or mounted outside near a rear perimeter door.

There are several different ways that these key storage boxes can be attacked, sometimes you can guess the combinations, sometimes you can decode a physical combination with a tool or just with your hands, sometimes you have to pick the keyway or find another hidden key, and sometimes (if you are positioned for it) you can use tools to attack the physical box itself.

[Below: Key storage boxes mounted to a portable plank of wood, used in our in-person Tactical Lock Picking courses when we run scenarios.
Photo Jan 2021, Denmark.]

[Opposite: same.]

CONCLUSION

Sometimes, it's worth a check of the Two-Step area before trying to pick or bypass a locked obstacle.

- When a strict range of "two steps" doesn't yield a hidden key, determine whether you want to search farther away or not.
- If you decide to search in an area outside of the Two-Step range, you should still be looking through the lens of common human patterns to aid you in your search for hidden or stored keys/codes.
- If your current plan is likely to make entry in less than 60 seconds, I would skip the hidden key search for now.
- You can always send extra people, if you have them, on the hunt for nearby keys and codes.
- If your attack timeline starts to look like it will take more than 10-15 minutes, or worse, more than an hour or so for something like a brute force code attack, then spending five minutes looking for a hidden key or code *first* would likely be a great option to add into your order of operations.
- If the locked obstacle that you're dealing with feels like a great candidate for having a hidden key nearby, check for a few seconds before your entry attempt. If the environment looks clean and sterile and there are little to no hiding spots for a key, maybe exhaust your TLP entry attempts first before the search.
- If you are going to store a key or a code for one of your locks, make sure you take a few extra steps to ensure that it is not easily found using the information provided in this chapter.
- Your Target Assessment phase can have many pauses in it. It is allowed to happen out of order. If you get stuck or frustrated, the Target Assessment steps laid out in this book can help you to get back on track.
- Go try. Go be willing to fail. Go try to troubleshoot in real time. Keep your eyes open for patterns that emerge through your own case studies. And keep in touch with us if you have a unique case study that you think we'd like!

[Distance between barracks room and kitchen stove: zero steps, just a pivot.]

[From a young age humans lear to hide things above eye level... and it sticks.]

[Front left or front right corner. Always a safe bet to start your search there.]

[Some of our key storage boxes that we give students practice on in our courses. Also pictured: the UTAC exclusive “Brute Force Notebook”, another tool to help you attack some common combination locks.]

CASE STUDY:
THE LIEUTENANT'S OFFICE

"Hey Dispatch..." I heard my lieutenant's familiar voice coming through my in-car radio.

"Go ahead, Lieutenant," Dispatch responded.

"Yeah, Dispatch, go ahead and take Three-William-Sixty-One out of service and send him to my office at District 2."

A screen popped up on my in-car laptop that said "Out of service - Location: D2." Dispatch then asked me over the radio "3W61 Do you copy?"

"Yup. I'm already 51 in that direction," I replied.

Most of the time when I was asked by another unit to be taken out of service to go take a special assignment, it was usually to help by interpreting American Sign Language, or to help someone gain entry somewhere (locker room lockers, squad cars with the keys locked in them while still running and, yes, on real emergencies too). Having no context from the LT, I was hoping that this was the purpose for the assignment, but it was a shot drive and I figured that I'd find out soon enough.

Being called into your boss's boss's office without any context certainly wasn't nerve wracking. I definitely wasn't pulling every mental file from every call that I'd worked over the last year. My heartrate definitely wasn't above its resting average...
...at all.

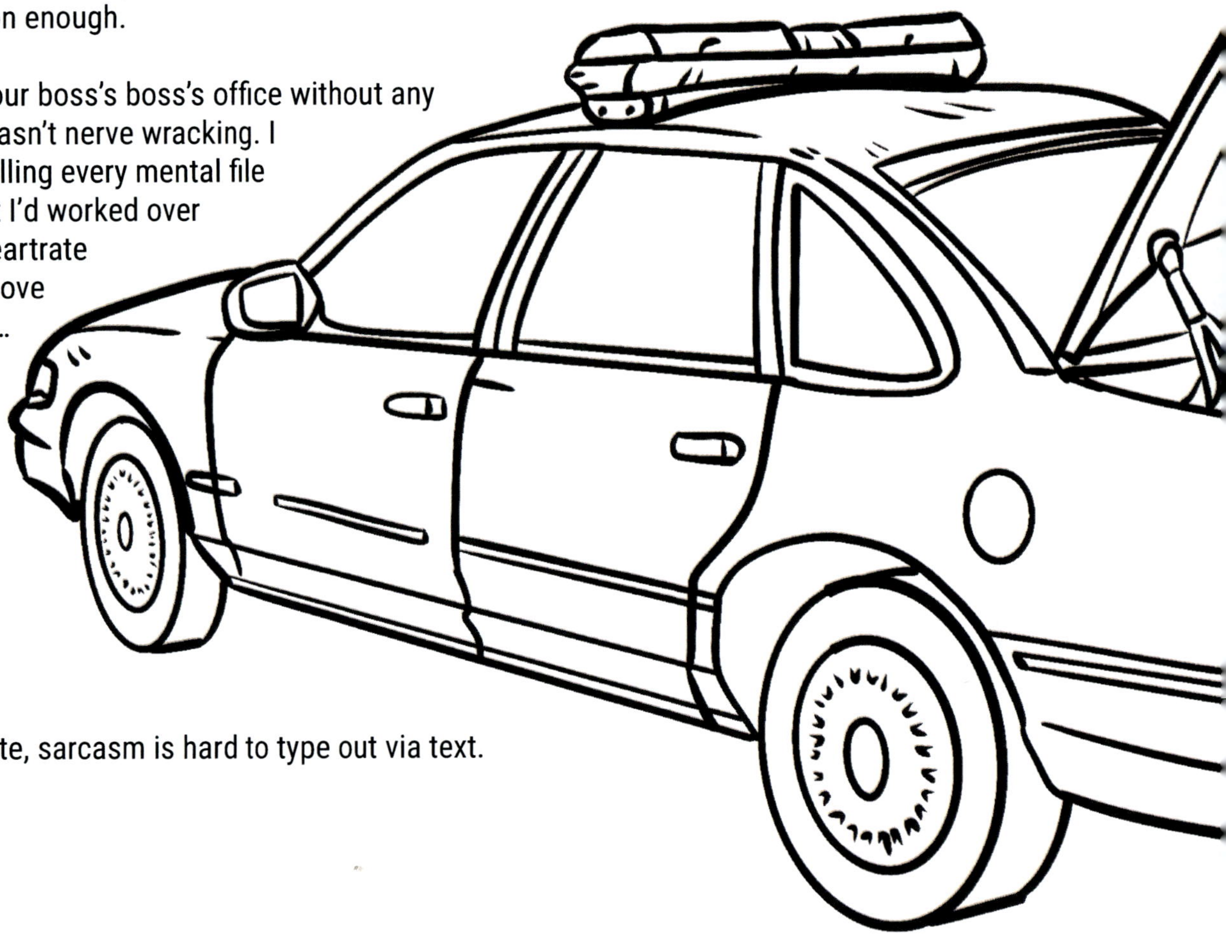

On an unrelated note, sarcasm is hard to type out via text.

I drove to the D2 building, reached up and tapped my garage door opener that opens up the employee-only gate, I drove in, parked, took a deep breath, and held my ID card up to the RFID reader next to the deputy entrance, then walked in.

Down the long single hallway that runs the length of the small building, I immediately saw the lieutenant and the Sergeant (the same one from the case study "Suidice by Pills") both standing in the hallway and both leaning forward and staring at the lieutenant's closed door with their thinking-faces on.

"Hey, gentlemen! Can I give you a hand with something?" I asked.

The lieutenant turned and looked at me and embarrassingly smiled "Do you happen to have your tools with you?" he asked.

"Are you kidding me? Of course I do!" I added: "This scenario is much better than the possibilities that I was considering on the whole drive here." I laughed a little, allowed my blood pressure to come down a little, and I started my Target Assessment.

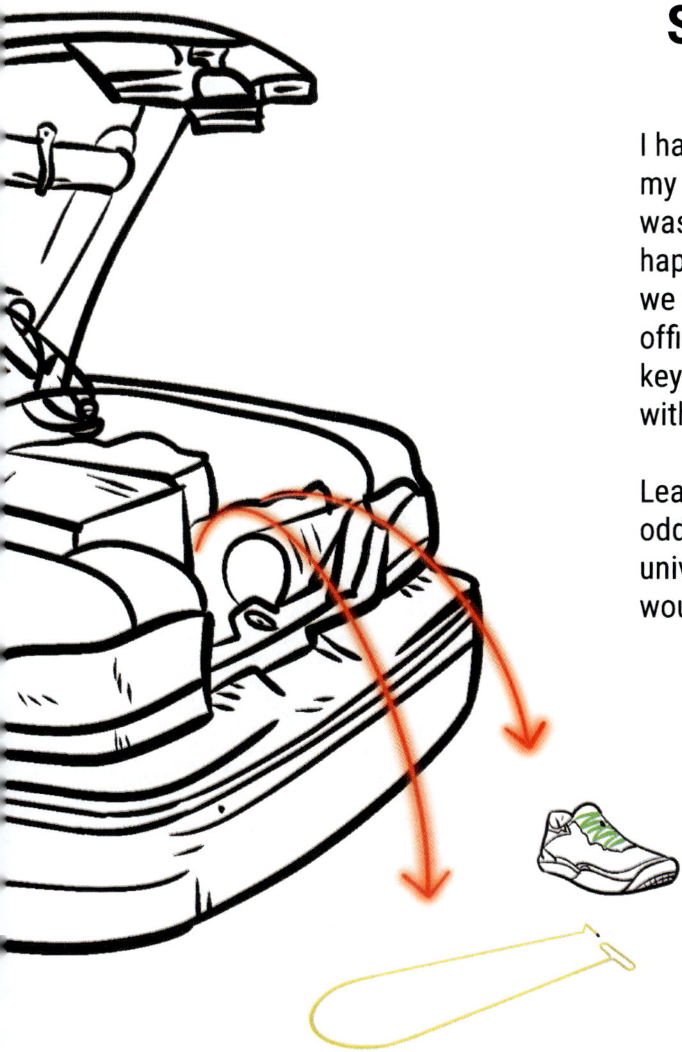

STEP 1 & 2: MRE AND SPEED

I had a clear Moral Right for Entry from the lieutenant, and my Level of Urgency was almost zero, for now. But the LOU was subject to change very rapidly. If a member of the public happened to have an "Oh shit! I need help right now!" moment, we would need to make very different entry plans into the LT's office. Those plans would be complicated by the fact that the keys to the lieutenant's patrol car were locked in his office, along with his gun belt and his bulletproof vest.

Learning those items were locked in the office, I thought, "That's odd..." (a phrase to pay attention to because it's usually when the universe is speaking to us). Aloud I said, "Hey, Lieutenant, what would make you leave your keys and your gun b-"

My face must have betrayed my thoughts because he interrupted me: "Because, I almost shit my pants running to the bathroom down the hall. Is that what you wanted to hear?"

I almost died laughing!

STEP 3: ATTACK VECTOR IDENTIFICATION

I continued with a quick assessment of the door to identify my Attack Vector options.

The door was solid wood. The frame was solid metal. So if worse came to worst, we would probably need a lot more than a swift kick to open it. Also, it was a push door with a commercial-grade keyway, a handle (not a circular knob), and no deadbolt, and under the door there was a gap big enough to drive a Matchbox car through. I knew without doing a full perimeter check, there was only one way in and one way out of the LT's office. The office had four walls, no windows to the outside, no other interior doors, a floor, and a hard ceiling (there were no roof tiles that we could move to climb up and over the walls into the office).

My two primary options were (1) to attack the keyway and (2) to attack the latch. I had tools for those options with me, and I had additional tools if I needed them in my Second Line Gear in my vehicle. But there was a third option too that I'd already identified as the most likely candidate.

A very important part of your skill set is knowing when not to waste your time lock picking. Rule of thumb for beginners, "Keyways with an extra squiggly, zig-zag keyway and/or a specialty icon stamp on the face indicates (usually, not always) that your pocket pick set is probably not enough." Especially if all you have in your First Line Gear are your small-sized tools. There are plenty of people that can pick open these medium- and high-grade security keyways with the right gear and a lot of practice and experience. I'm not saying that you should default to quitting before you start but I am saying that you should know the limits of your skill level and tool level so that you use your energy and time in the most productive ways during an entry.

[Above: the door we will reference for this case study. Solid heavy wood door, solid metal frame. handle on the left, door is a push-door.]

[Above: not a crazy difficult keyway, but I knew from looking at the Sgt's keys for his office nearby, that this keyway was going to be a little tough to pick.]

STEP 4: MANUAL UNLOCK CHECK (X2)

I went ahead with what should almost always be the first physical act in your on-scene Target Assessment: Manually try the doorknob (or whatever other obstacle is present), take a deep breath, then try again harder (thus, the X2). I keep repeating this for a reason! If you are sick of hearing it, that means that it is working its way into your head. The intended goal.

It might be insulting to a client but go ahead and double-check all locked options to confirm they're actually locked. And try again to open them...harder. If you're teaching others, remind them that it shouldn't be taken as an insult that you check their work, but it should be taken as an insult if you *don't* reassess their progress. There is a similar and common phenomenon in law enforcement and in some military operations: "Hey! Why are you searching that suspect for weapons again? I already searched him!" is something I've heard often in my career. Blame it on me if you want: our TLP curriculum states that you MUST try the handle, even if 10 other people checked it before you got there.

STEPS 5 & 6: TOOL AVAILABILITY & "NOW OR LATER"

I had my First Line Gear on me, so I took out my pocket sized rake and tension wrench. Since I had the time, I figured I'd try to make progress within the commercial lock's keyway. I drew a Line in the Sand: I wouldn't spend any more than five minutes trying to pick it open. I already knew what my second and third Attack Vectors would be, and I knew exactly when I would be switching to those attempts if my keyway attack didn't work. So I chose "Now" to put my Target Assessment on pause and to start a physical entry attempt.

I chatted with the lieutenant and the sergeant while I tried raking and picking. I explained a little bit about how pin tumblers work and I answered some of their questions about how lock picking in general works. This was me in my element! I could not have been happier.

The common usage of the phrase "lock picking" can mean many things. As a day one lock picking student, it usually means one of two things: either (1) picking open a lock with a raking technique or (2) picking open a lock using a Single Pin Picking technique (SPP). For this keyway, I quickly switched from raking to SPP since raking is most effective with low-quality locks. At the time I'd started practicing SPP more and could tell I wasn't making much progress. (Another example of something that you'll have to learn to feel as your skill set progresses.)

After about three or four minutes of both raking and SPP, I realized that I didn't have the right tools on hand, and/or the skill level to be able to pick open this keyway that was slightly higher in security than a standard residential front door lock. I had hit my limit - my Line in the Sand - for this particular Attack Vector.

Something I've added to the Tactical Lock Picking curriculum is the Line in the Sand "tool". Once you place your Target Assessment on pause and begin your physical entry attempt with a specific Attack Vector, you should start a timer (if not a physical timer, at least a mental one) for how long you will continue before stepping back and considering other options.

This prevents the all-too-common "Just give me a few more minutes," which turns into more and more and more. (You may have noticed in the case study "Suicide by Pills" that I had not paused and done a reassessment quick enough before the sergeant found the emergency key, sometimes a very short Line in the Sand is a good idea, and sometimes you can set aside a greater amount of time for a mental or physical timer before you pause and reassess.)

STEP 7: OTHER RESOURCES & RESET BREATH

Because the sergeant had given me the key to his office, I was able to go inside the sergeant's office and shut the door from the inside. I could see that there was a dead-latch mechanism, and unfortunately for me (and for the lieutenant), it was installed properly, which meant the lieutenant's latch was probably also installed correctly (more on this concept in "Same Lock Same Key" guideline chapter). That eliminated the possibility of using a shim on the latch since the purpose of a dead-latch mechanism (when installed correctly) is to prevent the latch from being manipulated.

At that time, (roughly 2014) I'd recently become aware of the under-door tool (UDT) but hadn't yet been able to purchase the tool and add it to my gear loadout. I checked the sergeant's unlocked door to see whether a UDT attack would even work or not since a UDT attack required a specific type of locking action in order to work. The UDT technique doesn't work on dead handles, only on alive handles.

[Right: A full inspection, inside and out, of the sergeant's door which was right across the hall from the lieutenant's office. Doing this is a textbook example of leveraging the "Same Lock Same Key" guideline. Often, *not always*, similar locks and keys nearby your target lock, or makes/models of locks and keys like your target lock, can tell you a lot about your target lock.]

Alive handle defined: This type of handle is commonly found on hotel room doors. If you're inside a hotel room, even if you lock the deadbolt, usually all you have to do when you want to leave is turn the door handle and the locks and the latches automatically retract so you can pull the door open. This is an alive handle, meaning that even when a door is locked, the act of turning the handle from the inside will automatically unlock it.

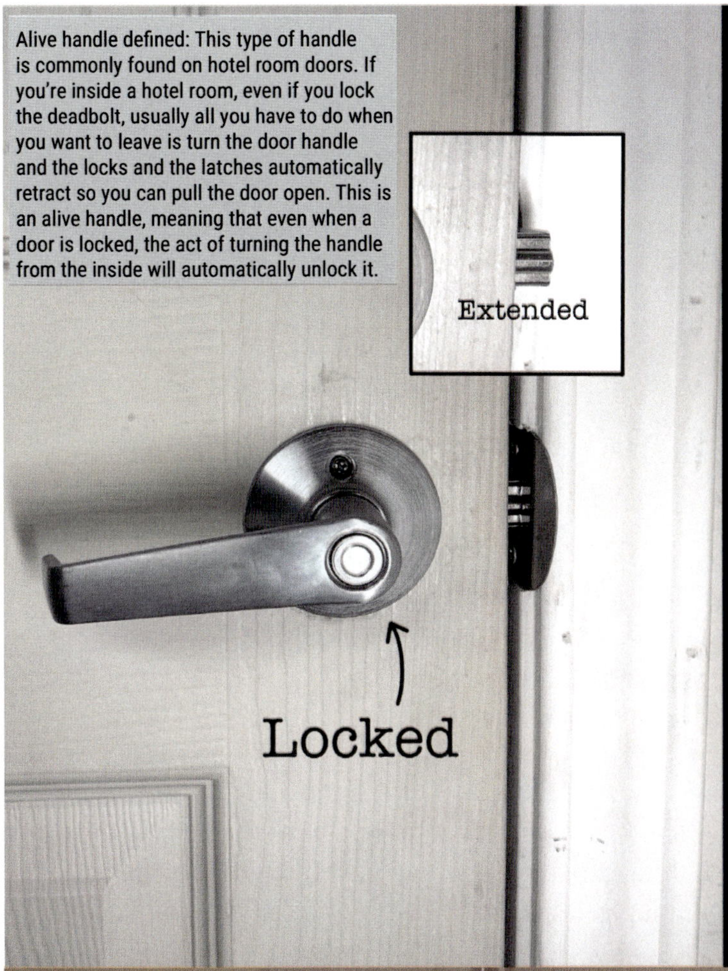

Dead handle defined: Imagine you're in a bathroom in someone's home and the bathroom door has a handle instead of a knob. You shut the door behind you and activate the twist-lock to secure the door. Once you're ready to leave, you turn the handle without thinking but the handle doesn't budge. When you have to manually disengage a lock to get the handle to turn, it's a dead handle: one that doesn't move unless manually unlocked.

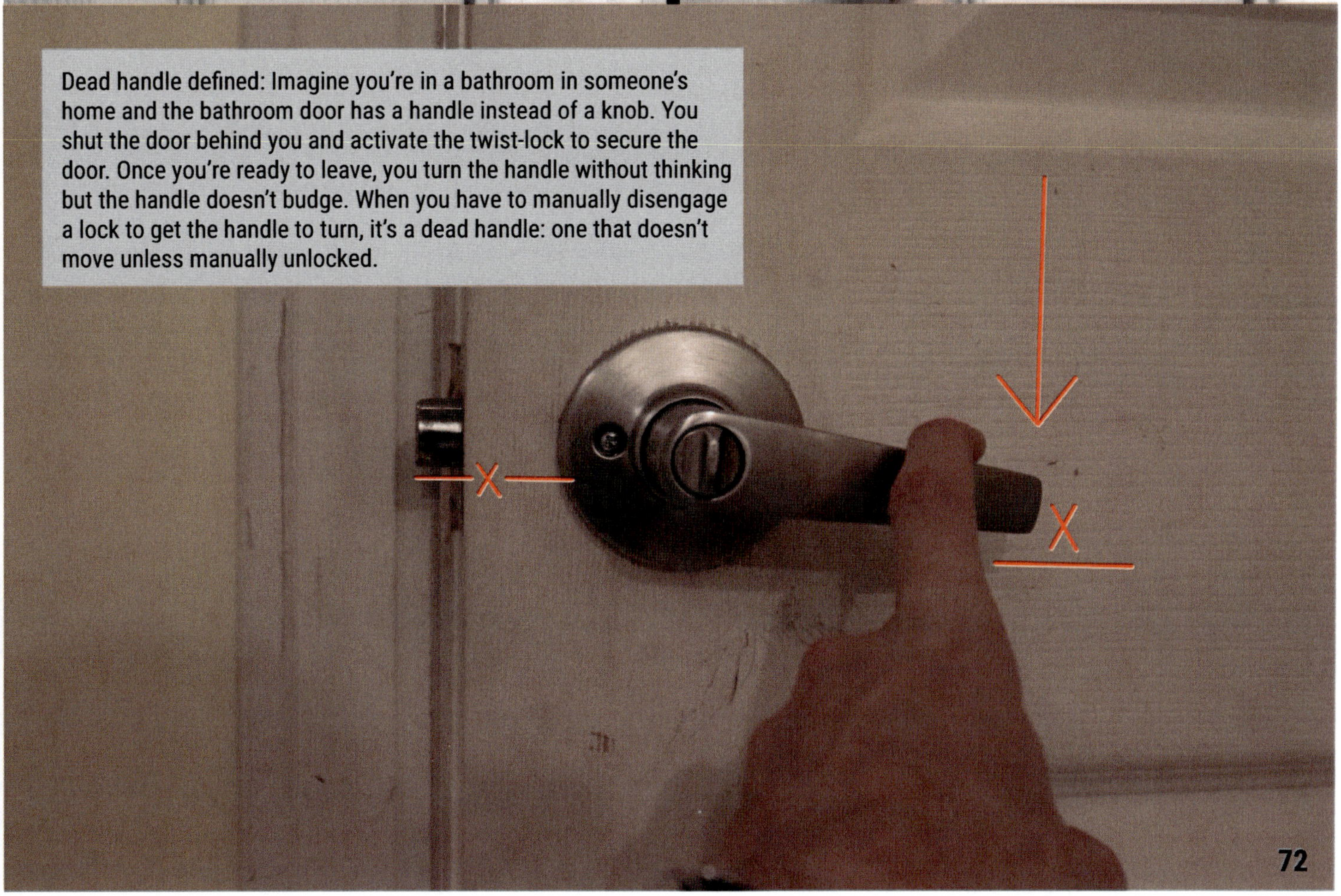

I did not yet own a commercial/professional UDT, but I did have another long, thin, bendable reach tool: A vehicle-opening device known as the Gold Finger tool. I love that name. I also had some string in the form of shoe laces from an extra pair of tennis shoes that I kept in the trunk of my patrol car. It was now another chance to prove the concept "Software over Hardware" - understanding a depth of knowledge about your craft is often way more important than having the exact most perfect lineup of tools at your disposal.

I grabbed the Gold Finger tool and one tennis shoe from the trunk of my patrol car and walked back into the building to the end of the hall where the lieutenant was waiting by his office door. A small part of my brain was thinking "If this doesn't work, I'm going to look like an idiot," but a large part of my brain was thinking "If this works, I'm gonna look freaking awesome."

Right on cue the lieutenant grinned and asked, "How the hell are you going to open my door with a fishing pole and a shoe?"

Leaning on the "Same Lock Same Key" principle - recognizing patterns from similar makes/models - I was able to bend my Gold Finger tool to the specific necessary shape, tie the shoelace to the top end of the J-shaped tool, and do a practice run on the sergeant's unlocked office door, with the door cracked halfway open so that I could cheat and peak my head around the door to watch my progress with the tool in real time. After that short practice run I had one successful run under my belt and all I had to do was replicate the exact action I had just taken on the lieutenant's door.

Excitedly, I knelt down, slid the J-shaped metal rod under the lieutenant's door, tilted it up toward the handle on the far side, looped the tennis shoe string around the interior door handle, and with a pull of the string, the handle turned and the latch receded into the unlocked and open position. Holding that J-shaped tool steady, and keeping a tight hold of the tennis shoe string under constant tension, I used my third-eye and leaned forward into the door. Viola!

From sliding my makeshift tool under the door to opening the door took about all of 10 seconds. Entry complete.

A bonus Return on Investment (ROI) to having this skill set is that you gain the ability to help others from all walks of life, resulting not only in helping others save time, save money, and save lives, but to gain "social capital" in the act of service to others during emergencies. If you are a "useful human", it becomes a little more difficult for supervisors to fire or demote you, and it also makes it easier for others to exchange value with you when they are able to.

CONCLUSION

I later spent time learning how to identify on sight which types of commercial locks are a higher level of security. This was an additional resource in the field that would help me manage my time and energy in the most efficient manner possible.

I added a reminder to my Target Assessment file to keep my eyes - and my mind - open for tools that I could makeshift out of my environment. (Since this case study which happened sometime around 2015, I have had to make and have successfully used makeshift UDT tools a fairly large amount of times. And every time it is a testament to Software over Hardware!)

If there are similar make/model doors, locks, etc., that are in our environment or that we remember from other case studies or from our research, we can use that information ("Fourth Line Gear") to our advantage.

I went ahead and pulled the trigger on my online shopping card for a professional UDT tool, it has been in my "Top Three Favorite Tools" list since that moment almost a decade ago.

Under a different LOU, especially with having a professional UDT at all times in my Third Line Gear, I might have skipped raking and picking, and might have skipped a latch slip attempt, and gone straight for my professional UDT. There's not always a wrong answer for how TLP practitioners flow through their order of operations on an entry, but as long as you know why you are choosing a certain route, that's the most important thing to manage the application of your craft.

TLP is a game of time management and of percentages. Even at a low level of practice you should have a fundamental, well-rounded skill set and be able to assess multiple Attack Vectors for many different types of locked obstacles. Increasing your abilities with different tools and techniques will increase your percentage of likelihood that you'll be able to open certain obstacles, but it will also change the way you navigate your order of operations on scene. For TLP, the real magic is HOW you make decisions to apply your craft, not just the usage of the tools and techniques.

SKILL: UNDER DOOR TOOL

This tool is a great representation of the way that lock picking is both misunderstood and often overlooked by wave of the hand assumptions.

The tool is "simple" in technique concept but it's not always "easy". This is starkly different than what you will see in post after post on social media. Almost every UDT post that you'll find is people carrying out a successful entry from start to finish, smooth, without any struggle whatsoever.

Here are a few tips to help get you started.

- Normal usage: the tool and the pull-string go under the door, you position the tool around the handle on the far side of the door, you pull down, and the door swings open like magic!
- Sometimes there's no room under the door for this tool to fit. Non-starter.
- Sometimes there is only a tiny amount of room, and the small space below the door "pinches" the tool while in use so it really throws off your tactile feedback and really slows down or even causes an entry failure.
- Sometimes it is hard as hell to position your tool and the pull string around the far side latch, even on a seemingly simple door.
- This technique only works on "alive handles" not on "dead handles" and sometimes there's no way to tell which one you're up against (defined in the previous chapter).
- Sometimes the pull string wraps around the body of the tool, even with the tool is in the right position your pull string will not have captured the latch and you'll have to keep starting over until you realize your mistake.
- But when this works, it's really amazing.

[Opposite: UDT instruction block during our Advanced Covert Operations Course, 2022, hosted by ProtectiveOps. On a stubborn or tricky handle, there can be quite a bit of troubleshooting that has to take place, as well as a suite of complimentary tools to assist.]

UDT POSITIONING

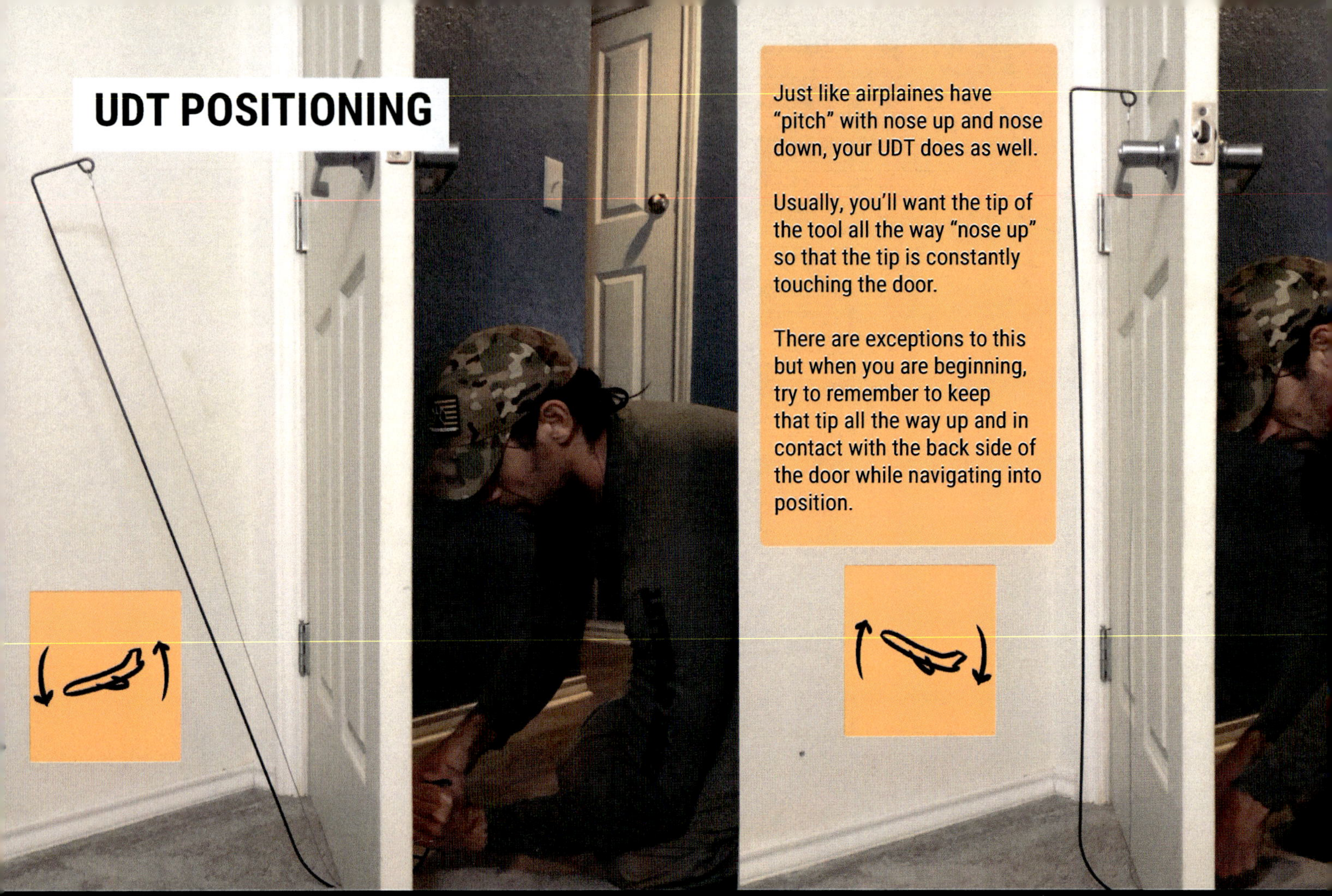

Just like airplaines have "pitch" with nose up and nose down, your UDT does as well.

Usually, you'll want the tip of the tool all the way "nose up" so that the tip is constantly touching the door.

There are exceptions to this but when you are beginning, try to remember to keep that tip all the way up and in contact with the back side of the door while navigating into position.

And like an airplane will roll left and right, our UDT too can roll left and right, like a windshield wiper. Even a small angle change with your grip can have a large effect on rool, and all of the other positions of your UDT.

Your track line, or positioning left to right, also like a car driving in the left lane or right lane, can help you to leverage certain advantages as well. Some people stay middle of the road and try to "roll" or tip the UDT toward the latch, others like to line up their track line right beneath the handle and pull straight down.

This is easily the most overlooked positioning issue that I see with students in our courses, the "yaw" or twisting/spinning left and right of the tool. A UDT operator must be acutely aware of all of the different positions of the tool on the far side of the door and must always remember that small adjustments on the tool's handle have large effects on the far end of the tool.

CASE STUDY: LOCKED IN

The sun was going down, the daytime animals were slowing down and the nocturnal animals were starting to perk up. The the sole Ranger at the park drove up the only access road towards the park's entrance/exit. They drove out and parked just past the double steel manual swing arm gate that was left wide open during park hours.

The Ranger walked to the right side of the road and unhitched the right-side gate from its open position and walked it closed across the right side of the street. They walked over to the left side of the street and unhitched the left-side gate and walked with it as they swung the heavy gate arm shut to the middle of the street, meeting with the right-side gate. There was a short chain with a padlock at each end, hanging from one of the gates, one padlock at the bottom of the short chain hung open and unlocked. The Ranger looped the open padlock around the opposite gate, locking the gates shut for the night.

There was one patron left in the park

"Three-William-Sixty-One, I'm attaching you to a call in your zone at **9591 State Park Road**. A guest at the state park just called in, they're locked in behind the park's gate and the Ranger has left for the night. Can you be 51?" Dispatch informed me on the radio.

"Yup, I'm already headed that way, thanks Dispatch," I replied.

Another call specifically dedicated to a locked problem, rather than a call having to fight people or argue with people? My lucky day!

I curved my way down the long scenic road as dusk grew into dark. The last curve towards the state park revealed a pair of headlights behind a locked swing arm gate.

I parked. I hit the "arrived" button on my in-car laptop, and I started chatting with the locked-in park patron.

Here's what he told me, "I had no idea that the park was closing. Nobody told me, and there were no signs posted anywhere. I just walked back to my park after hiking and found this ticket stuck to my windshield that said 'You can retrieve your vehicle Monday-Friday during open park hours for a $15 fee' and I just don't feel like that's right. There was no way for me to know this."

His story seemed genuine, and I believed him. We were, however, talking across a closed and locked gate and when I checked his ID we actually both reached over the gate right above the sign that did in fact read "PARK GROUNDS OPEN SUNRISE TO SUNSET."

There was a little blip in my head that said "Are you SURE there was no way for you to know the park hours?" but I am also a reader of history (government) and of systems design. Something told me to put that thought on hold, so I did. Instead of exploring the rabbit hole of the park rules and the ticket, I instead shifted gears to my entry (exit?) study.

The next few pages of this case study took place in the span of about 60 seconds...

CLOSED CIRCUIT
TELEVISION
AND
AUDIO MONITORING
ON PREMISES
PARK GROUNDS OPEN
SUNRISE TO SUNSET

[Left: From the steel swing arm gate on the left, to the steel swing arm gate on the right, we had "Keyed padlock - short links of chain - combo padlock".]

"JuSt uSe yOuR bOLt cUtTeRs" ...and leave the state park in your patrol zone wide open for the night, or over the long weekend? Hm. Seems that the people from the social media comments section may have spoken too soon on this one.

This is not a "lock picking" scenario. Meaning: this is not a test to see how a lock picking tool or technique can be forced as a solution to a problem involving locks. We could have told the park patron "Good luck" and drove away. We could have got in touch with the off duty Ranger and had him come back to the park. We could have given the patron a ride home. I could have smashed through the gate with my patrol car.

You may have just caught on that "can doesn't always mean should". This was a scenario being played out in real life, with real people, in a real environment, so, it was not a "scenario" on track that is designed to progress using a limited scope of tools/techniques like in a classroom. This was a full spectrum, real life problem. We had to make decisions that were reasonable within the totality of the information we had at hand. If lock picking was the best answer, we'd use that. If bolt cutters were the best answer (they were not) we would have used them.

Remember, at UTAC we don't teach exclusive curriculum, closed curriculum that is limited and unchanging. We give you foundations that you get to adjust and alter and to add into your toolbox in whatever way you wish. We do not claim that Tactical Lock Picking should replace explosive breaching or ballistic breaching or even mechanical breaching for things like bolt cutters or battering rams. This systemized approach is meant to be added in as an option for your real life toolbox.

For this scenario I decided to see if I could employ my craft: the systemized approach of lock picking and other non-destructive entries.

It took all of about 4 seconds to grab each of the padlocks on the chain for a quick peek at their keyways and combos and other make/model info. I was familiar with both types of lock as well as some tool/technique options and what the difficulty would be that I could expect. I also, immediately, had a first, second, and third option for my order of operations, or which methods of entry I'd try and in which order if they were unsuccessful.

Pause for a moment, if you are familiar with either of these types of locks (or even if you're not), take a second and ask yourself what tools or techniques you might try to gain non-destructive and destructive entry and in what order.

Padlock #1 - A "BEST" brand padlock. I knew their keyways to be outside of my skill level but if I absolutely had to, I could try to get lucky and see if my picking would somehow gain me an entry tonight, but this was one of my last options.

Padlock #2 - A "Master 175" combo lock. I knew at a glance that there were three separate attack vectors at hand that I had for this particular lock, two of them were at the top of my list, and the other one was even below picking the best lock as my third option for this lock could have taken a loooong time.

After the first four seconds of this entry case study, I let go of the padlock in my hand and I let it swing, and I told the park patron "I'll have you out of this park in about 30 seconds, and you won't have to pay for that ticket either."

I wanted to get the most out of this case study, and I had a hunch, so I added another layer of data to pull "Hey Dispatch, can you please check the code book for this address and let me know what the four-digit code is for this padlock here at the state park at 9591 State Park Road?"

"10-4 standby," Dispatch responded. Another mental timer started, tik-tok, tik-tok.

I went to the back of my patrol cruiser and grabbed my Second Line Gear bag from the trunk and pulled out a tiny metal tool called an EZ Decoder (update: I now carry two of these inside my wallet, my First Line Gear, 365 days a year).

I held the Master 175 in my left hand, and with the EZ Decoder pinched in my right hand, I inserted the tool into the third wheel to the right, in the top left corner of the wheel slot.

I held pressure down on the shackle with my left hand thumb, I rocked the tip of the EZ Decoder up with my right hand, and I released the pressure on the shackle with my left hand and the padlock popped open with that classic ::ka-chink:: sound.

A few more seconds passed where I walked one of the swing arm gates open, waved as the patron drove out, then I closed and locked the gate. With the gates locked and the bottom of the combination lock was still showing "9-5-8-0," I got back in my cruiser and wrote some notes into my call log using my in-car laptop, and I opened a running notepad file of lock picking case studies that I managed to pick up on my shifts.

"Three-William-Sixty-One?" Dispatch asked for my attention on the radio just as I was typing up a short lock picking case study titled "9591 State Park Road - Locked In."

"Go'ahead," I responded as I looked over at the address in large 10-inch tall text attached to the park's front entrance sign. "...that combo is going to read **nine**... " a short radio pause "**five**..." a short pause that I interrupted.

"**Nine-one**?" I said on the radio. To which dispatch confirmed.

CONCLUSION

- For the first padlock, the "BEST" padlock, there are now pin-tumbler Lishi tools designed for this keyway. That could have moved a keyway attack on the best lock up a little bit in my list, or at least could have increased the percentage of likely success for when I did engage in that attack vector.

- There are a lot of things you can do with four-digit codes that we will discuss later in this book. One of the simplest ways you can attack a numbered code is counting from 0-0-0-0, 0-0-0-1, 0-0-0-2, etc. This can take quite a long time, or can be surprisingly short. This "counting attack" is called a "brute force" attack. The brute force of trying all possible combos, not a brute physical force.

- Especially on chains, more locks doesn't always mean more secure. This is also seen on doors and other types of obstacles as well. In this case study, if we had accessed *either* padlock, we would have gained full access. Sometimes you have to unlock *multiple* locks to gain access. Keep this in mind when assessing a target.

- We didn't dive too deep on all of the separate Target Assessment steps in this case study since the stakes were low, the timeline ended up being pretty short, and I knew of three or four likely methods right off the bat that could gain us entry, and I just decided to immediately go hands on with one of the sure-shot methods in my toolbox. There is no reason that you MUST follow every step to its conclusion, in order. If you walk up to an obstacle and know a pretty good technique, there's nothing in our curriculum saying that you CAN'T just go try that thing first.

- There isn't a ton to learn from this case study because, as we've said before, starting with our Dedication page, you can learn a lot more from your struggles and your failures than you can from your perfect successes.

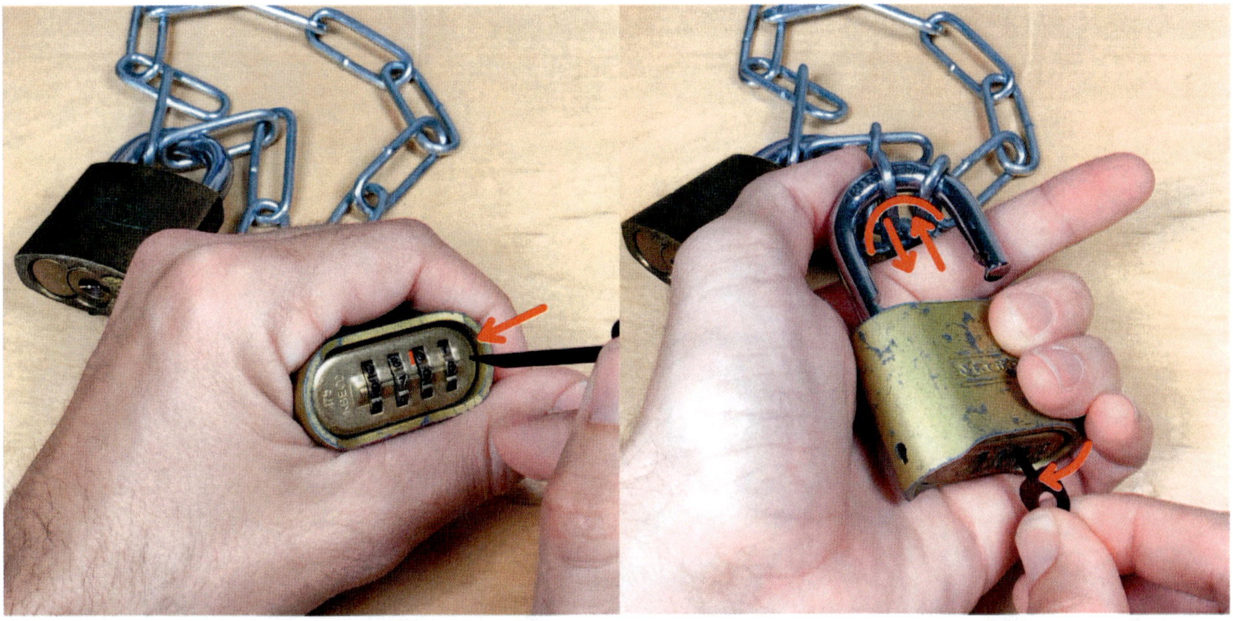

[Above: When this bypass technique is applied without any troubleshooting required, it can often take as little as one-to-two seconds from the insertion of the tool to the opening of the lock.]

- The addition of calling dispatch and asking for the code was an attempt at a race of sorts. It was designed to add one small tick to the support column of proving the phrase "Sometimes, lock picking and other non-destructive entries can be faster than asking someone to look up the code for you." These types of entry methods have also proven time and again that they can occasionally be faster than some destructive breaching scenarios.
- There was in fact something fishy with the park hours sign. It was below eye level and proudly displayed on a *swinging* gate! The gate rests open all day thus making the signs on it not only very difficult to read but just difficult to even notice at all.
- And, we'll dig into this more in our four-digit code chapter, but do you remember the code that the combination padlock was showing when we peeked at it? And does that look really close to some other type of number that we may have encountered on this case study?

SKILL: MASTER 175 BYPASS

To apply this technique:

- First you need to identify that you're working with the right type of padlock. This technique works well on older "Master 175" combination brass (or brass colored) padlocks. Newer locks often have tighter tolerances and this technique **is not** 100% effective on 100% of the Master 175's out there. You'll see in the image above that "maybe" this will work on the combination padlocks shown above. But just because you are dealing with a padlock of the right "make" and "model" does still not guarantee that this technique will work. Sometimes there's a sub-model number, or a manufacturer's date that will produce padlocks with different internals.
- Next, you'll need to make sure you have the right tool with you, and/or you can makeshift one from your environment. This term "the *right* tool" is also subjective. Sometimes an EZ Decoder works best, sometimes an Ultra Decoder (both pictured above) will work better. Sometimes the lock has such loose tolerances that you can use the broken end of a hair barrette/snap-clip. And sometimes it's a skill level problem where you may not be able to open it but someone that tries it with a little bit of a different angle and pressure and timing can open it successfully. (The value in TLP is **how our brand of entry operators troubleshoot**!)

- [Left: These locks look similar but the "Master 175" bypass technique in this chapter will NOT work on these locks. The internals are completely different from the locks on the opposite facing page.]

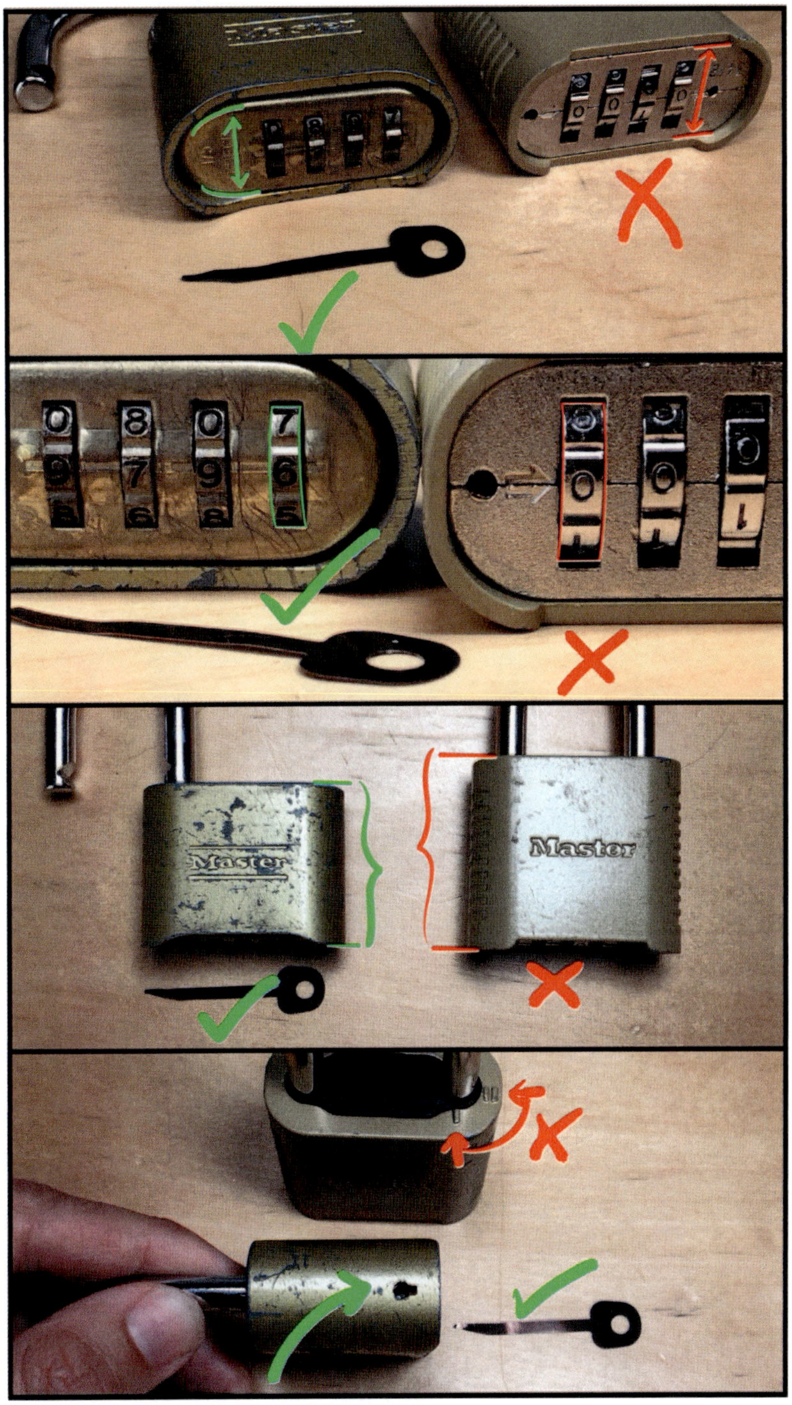

[Right: How to tell the difference, even when the general shape of the locks look similar.]

- Rounded brass bottom plate v silver flat bottom plate.

- Small brass wheels v larger silver wheels.

- Smaller body v larger body.

- Side access hole for combo change tool, v "I" & "II" markings for combo change without tools.

While the bypass in this chapter wont work on the larger "change your own combo without a tool" locks, we will teach you a decoding method for them later.

With the right lock type and the right tool, insert the thin metal tool into the lock in the third wheel cutout, in the top left corner alongside the wheel. It *shouldn't* really matter which numbers the wheels are currently set to, but if you're troubleshooting, there's nothing wrong with giving the wheels a little spin.

Sometimes it's a tight fit, sometimes it slides right in. Sometimes the third hole is too tight for whatever reason and it's worth experimenting putting your tool into different holes. Usually though, you'll have success with the third wheel-hole first, as a default.

Make sure that you have the numbers reading properly, right-side up, not upside down. this will not work the same way if you invert the lock and end up leveraging you tool in the wrong direction.

For depth, the tool only has to go in about one inch or so. Hardly ever will you need more than that for this technique.

And, with the numbers reading right side up, you'll want the tool to insert tip-angled-down, and not flat or aimed upward.

Once the tool is properly in place:

1. [Left: Press the shackle dowmn and hold.]

2. [Right: Rock the handle of your tool down (tip of your tool up), and hold.]

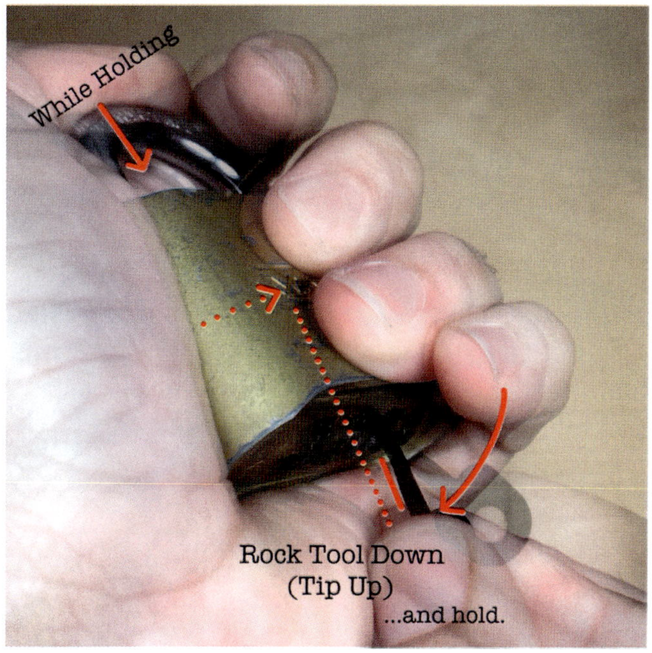

3. [Lelf: While holding the base of your tool down, release pressure on the shackle.]

CONCLUSION

Everything in TLP is a precentage game of likelihood and a percentage of possible success. This technique included. Sometimes it works as effectively as magic, sometimes it just will not work. Use all of the content in this book and in our courses to help you plan, and to help you troubleshoot for when life intervenes.

CASE STUDY: EMBARRASSED

Deputy T and I both arrived at the same time to a late-night alarm in our zone. A motion detector alarm had gone off at a large construction business just off the main road in the the patrol zone that we were working together. We each parked in the business parking lot and walked towards the small office building. Everything looked and sounded secure so far.

Joined to the side of the building was a seven-foot-high fence with a padlock hanging on a thick chain where the two swinging segments of the gate met.

Deputy T was familiar with the owner of the business and had been given permission from the business owner to open the padlock on the gate if he ever had to look around the property at night. While Deputy T tried the combo, I couldn't help but thinking, "Man, this would be a great opportunity to showcase how effective a lock picking or non-destructive entry could be!"

My wish was granted. Deputy T was having a hard time remembering the four-digit code. I excitedly rubbed my hands together in preparation for what was going to be an open and shut case study (pun intended).
"Hey, T, check this out," I said.

Knowing there were several makes/models of four-digit combo locks that were susceptible to the "Master 175" bypass technique you saw in the last chapter, I grabbed a hair barrette from the glovebox of my car (this is an old case study, I now keep tools like this in my EDC). I proceeded to explain to Deputy T how you can snap a hair barrette in half at the right place and how you can manipulate one of the levers inside this combination lock so that it pops right open.

I rotated the combo lock upward with one hand, and with my other hand I started the process of inserting the snapped hair barrette into the third wheel, in the top left of the wheel. I did not get very far.

I said out loud, "Hmm. Third wheel. Top left..."

I took a deep breath, and I tried again.

After a few more awkward attempts at troubleshooting, Deputy T said, sincerely, "Hey man, I'm sure that when that works, it's pretty nifty. I'll just take a stroll around the outside perimeter of the fence and clear the call if everything looks 10-4." He patted me on the shoulder and started off on his walkaround of the property.

There were no signs of anything being disturbed. If I had a reason to stay there, and a reason to access this gate, there were a large number of things I could have done next to approach this lock different with different Attack Vectors (covered later in this book), starting with trying the last four digits of the address, or the last four of the business phone number, etc.

Deputy T made it back from his perimeter walk and keyed up his radio "Dispatch, all clear here," and I blew out a deep breath in frustration (and a little embarrassment).

CONCLUSION

Always choose an entry method that is reasonable for your specific circumstances. For us, we didn't *need* to make *any* entry. If we were suspicious of something, and if there was no emergency, we could have sat there all night with one of us watching over the fence line and the other trying a brute force attack on the lock (a counting attack). Or we could have climbed the fence, or used bolt cutters, or tried to call and wake the business owner. Sky's the limit. Remember that the use of this **filed** skill set does not exist in a vacuum. You should not try to force any of these techniques just because you know them and you practice this skill.

After this failure, I learned that these types of locks are not able to be bypassed like the Master 175, although they look kind of similar. The internals are completely different. My research did, however, show me that there are two really cool, really effective and efficient entries that do work on this lock. Now, those techniques are often class favorites in our in-person courses (decoding each of the four wheels with a feeler-gauge, and decoding these locks by hand, without any tools).

In addition, any time you are shoving tools into a lock, there is a chance you could damage the lock or have a tool get stuck or broken or become non retrievable.

I was humbled to learn from this and I now remember to start every entry attempt with "This might work..." instead of "Hey, watch this amazing accomplishment that's about to happen!"

I now always carry tools for this lock (and the Master 175, same tools, "EZ Decoders") in my First Line Gear.

This call took place almost 10 years ago. And this failure was a very important lesson that led to a very important part of the development of our TLP curriculum ("Same Lock Same Key" chapter).

The value of Tactical Lock Picking is not in the selection of tools, or the selection of techniques that we present to our students.

The value is in preparing students to make successful field entries, even when several little failures happen during the entry process.

The - value - is - in - navigating - the - troubleshooting!

Software over hardware. Applying a system of approach is better than simply buying a bunch of tools that will *hopefully* work when you try to use them.

[Opposite: teaching students in our Tactical Lock Picking Course, in Wilmington NC, about the internals of this lock series and how it is different than the "Master 175".]

SKILL:
COMBO LOCK FEELER-GAUGE DECODING

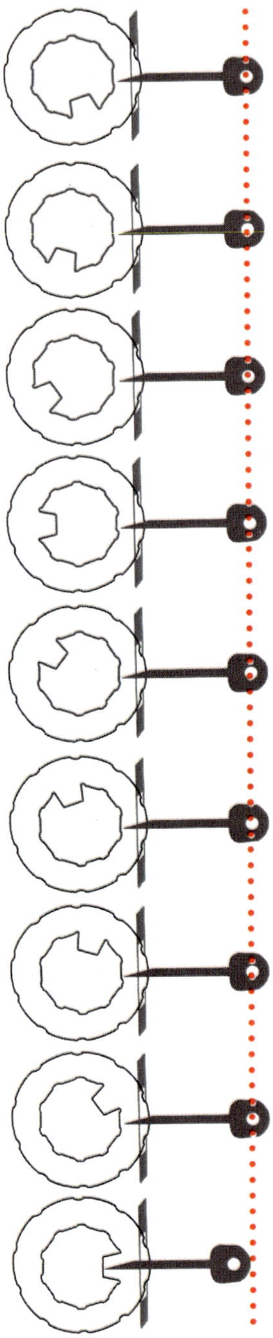

There are many ways to bypass a combo lock, or guess the code, or otherwise gain access to a combination lock. Many types of locks fall under the term "combo lock". This skill lesson is dedicated to physical (not digital) combo locks with multiple separate wheels, and the technique of feeling for the physical gates connected to each wheel.

Step one: if you have the right type of locking mechanism in front of you, and a tool that is able to work for this technique, you'll start with just one wheel on the lock. Using just one wheel, insert your EZ Decoder, or similar tool, adjacent to the first combo wheel (on the left, or on the right of the wheel) and explore how the inner wheel feels for the current wheel position. Then, spin the wheel, adding one digit to the facing associated number, and feel again. Once you complete an exploratory feel for each of the 10 (or however many total) positions of the wheel, one of those positions should stand out as markedly "different" [like the artwork to the left].

That one different spot is probably where the "gate" of that particular combo wheel lies. If you've felt nothing, try going around the wheel again, feeling each position on the opposite side of the wheel. If you're unfamiliar with the type of lock you're dealing with, you may have to explore all ten positions on the left of the wheel, then switch and try to locate the gate on all ten positions on the right of the wheel until you find something that is definitely different.

[Above: a side view of a disassembled combination lock wheel. You can see the "gate" that you're looking for inside the lock, adjacent to the main wheel.]

Step two: repeat this "gate identification" process on all of the remaining wheels until all of the wheels have their gates located facing the same direction.

[Left: For this particular make/model lock, the gates you are looking for will be to the right of the wheel. This is not always the case for every make/model lock. Some gates are easier to locate by sticking your decoder tool straight into the space next to the wheel like a "depth check" and other times it is easier to locate the gate by "slicing" your tool across the inner wheel and trying to feel a physical and/or audible click.]

Step three: with all of the gates lined up facing the user, rotate all of the wheels 180-degrees so that gates are all facing inside the lock. Then attempt to open the shackle using whatever mechanism is provided (pulling the shackle, or manipulating a lever, etc.).

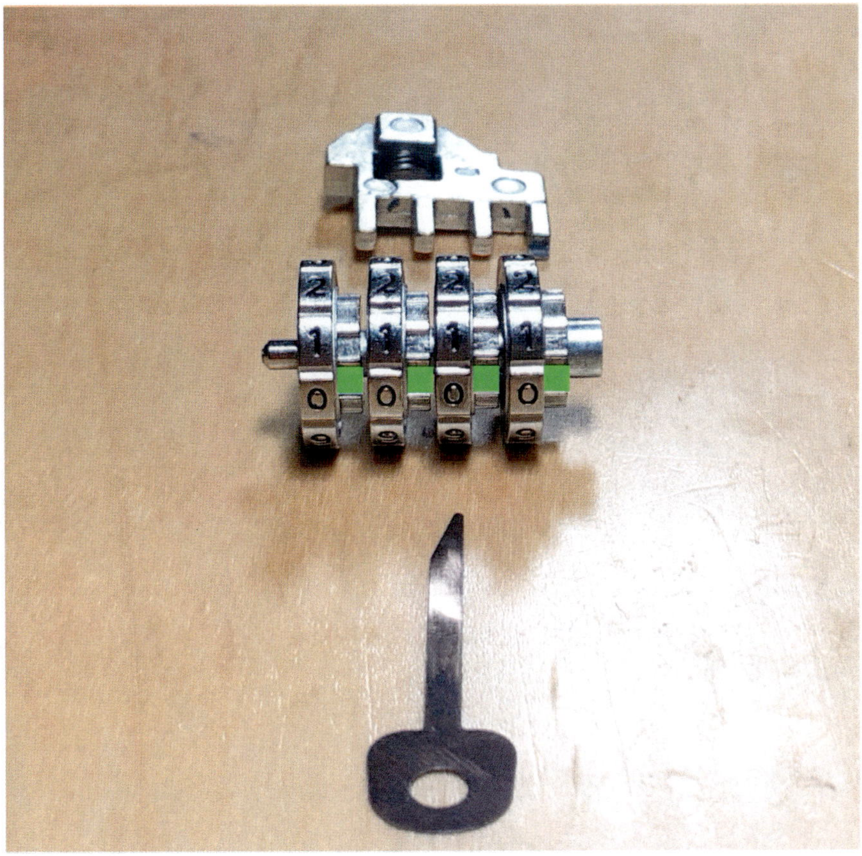

The reason for the 180-degree rotation sometimes throws people off. When using a decoder to feel for the gates, you are first reaching in from the **outside** of the lock to find and then align the gates... [Left]

...then, the gates need to be rotated towards the **interior** of the lock so that the mechanism [bottom of the photo to the right] has access to the opened up gates. [Right]

Troubleshooting: Sometimes the gates require a rotation that is different than 180-degrees. If you are sure that all the gates have been clearly identified and that they are all lined up together, facing the user, you can rotate all of them one position up, then try the lock. If the lock doesn't open, rotate the wheels one more digit up, and try the lock again. This will ensure that no matter what "offset" the gates need to be, you will have tried the gates in every possible position.

CONCLUSION

This technique is another example of "simple but not always easy". The concept boils down to: locate the gate on one wheel, then locate the gates on the remaining wheels, then rotate all the gates (together) to the correct offset. Doing this in the field is not always easy.

Sometimes there are mechanisms inside of the lock that block and prevent access from your decoder tool reaching the inner wheels. Sometimes those blockers can be bypassed, sometimes not. Sometimes you can do this technique by feel, applying pressure on an internal bar and then rotating and wiggling the wheels until you find a place in each wheel that "feels" different, even without a tool, just by hand. And sometimes you can apply this technique by sight if the inner wheels are easily visible through the face of the lock.

If you're just getting started, the best thing to do is to grab a few different multi-wheeled combo locks, and a few EZ Decoders, and to just start exploring all the positions of all of the wheels until you can say for certain that you have identified a pattern that exists where ONE spot on each wheel feels different than ALL of the other spots on the same wheel.

This technique is nice because it tells you the lock's proper combo code when you have successfully completed this type of entry. For Tactical Lock Picking, especially for our "emergency non-destructive entry" framing, figuring out the combo may not really be of any importance. But, for operational use, if you end up needing future access, or repeat access, or shared access through a combo lock, having the combo figured out could be a great help.

[Below: this decoding technique **does not** work on every make/model of multi-wheeled combo locks, but it does work on an awful lot! As of this writing, it **does** work on the brinks lock pictured below.]

CURRICULUM: SAME LOCK, SAME KEY

Curriculum guideline: using information from one lock to help you open another.

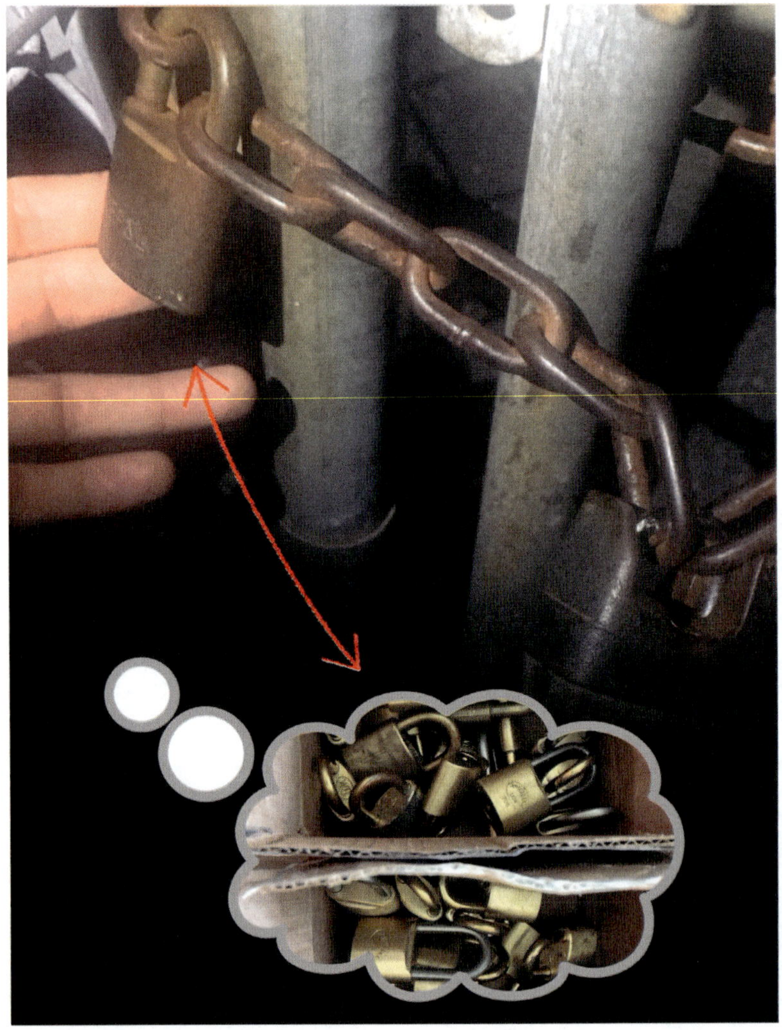

[Above: While inspecting a chain locked gate on scene, "Oh, "Oh! I recognize that type of lock! I have a box of those at home that I can open up almost every one of them in less than a minute each with the same lock pick set that I have here in my pocket." Not a 100% guarantee but a great tool to help guide your Target Assessment on scene. Same Lock, Same Key.]

- If you are looking at a lock during a lockout: "Oh! I recognize that type of lock! I have a box of those at home that I can open up almost every one of them in less than a minute each with the same lock pick set that I have here in my pocket." Same Lock, Same Key. There's a good chance (not 100%) that you can successfully gain access with the knowledge from those other locks. (As a common exception, sometimes the bitting pattern within a lock can have an effect and make entry easier or more difficult. Beware.)

- Multiple entry points: The key to most people's front door probably opens the deadbolt and the knob below it. Their front door key probably also opens a side or rear door to their house. Not 100%, but very commonly. In a large facility, most doors are "keyed alike" in some form or fashion, doors will be designed so that one key (or passcode) can open up multiple doors.

- Fleet vehicles: Vehicles that are ordered to all be openable with the same exact key for each vehicle upon delivery. This pattern is not uncommon. Sometimes even a key from a differently bitted lock will work in a lock that it's not supposed to work in, if it is cut close-enough to the proper key! It's not just building keys that can open multiple doors, it's for vehicles sometimes too!

- Entry timeline expectations: Same Lock, Same Key can give you a history of how long it may take you to open certain locks with certain tools and techniques. This can be leveraged during your Target Assessment phase "Hmmm, usually these types of combo locks can take me under 60 seconds, consistently, to open. But those keyways are different, sometimes 5 seconds to unlock and sometimes 15 minutes."

- Combo locks: Many combination locks leave the factory with the same preset code, or a manufacturer default code that is programmed into every lock. It is then up to the user to change the code. At this point, you may not be shocked to learn that even weapons, narcotics, and evidence rooms are secured behind a simple mechanical combination lock that still uses the manufacturer default code. (More on this in our "Four-Digit Code" chapter.)

- Possible techniques: Some specialized techniques that work on one lock will work on similar locks of different makes and models. SOMETIMES. Sometimes this works because of "clone" locks, where companies copy the exact same construction of another company's lock but they slap on a different name brand (or no name brand). Sometimes it works because the construction is similar enough. And sometimes it works because of dumb luck. But be aware that sometimes it doesn't work at all (as you saw in the chapter "Embarrassment").

- Common Construction: Construction companies save time and money by purchasing and installing identical door hardware on as many doors as possible within the same facility, and even within the same neighborhood or even separate buildings around the country but owned by the same company. This means that if you are picking or bypassing one door within a facility or business, you can probably expect the same percentage of likely success with other doors within the facility.

Case studies exist for gigantic agencies keying all their office suites the same in an entire city or an entire state; if one key goes missing (or is stolen) from one branch, then every single branch in a city or state is compromised. I teach an entire block of instruction on a lock that's utilized by an international package-delivery company. This company has 25,000 lock boxes throughout the US that use the exact same model of lock and the exact same push-button access code for every single lock box.

CONCLUSION

We can see that locks often speak to each other, or better yet, they have a relationship with one another that can speak to you if you know how to listen.

Be aware of exceptions to the Same Lock, Same Key approach to recognizing patterns, as well as for all of the guidelines, patterns, and curriculum within this book. Nothing is 100% effective.

Tactical Lock Picking and its application are always a percentage game: How low you can keep your expenditure of time, money, and effort, and how high you can raise your likelihood of a successful entry. Just as "Security is a balancing act between effectiveness and efficiency," we are constantly weighing our possible vectors of attack against how long they might take and how likely they are to succeed.

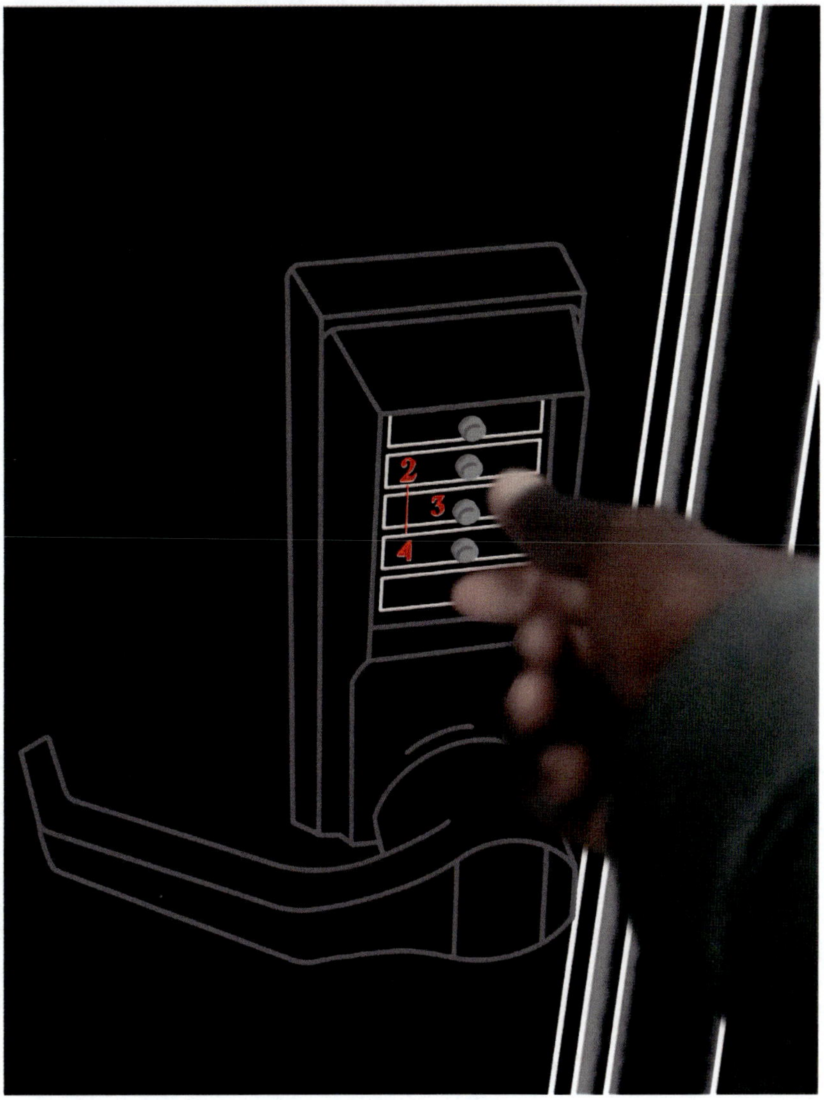

[Right: Government agencies, the U.S. military, thousands of local law enforcement agencies, hospitals, and many other companies use this exact make and model of lock and never change the code from the manufacturer's default: (2+4 at the same time, then 3).]

CURRICULUM: FOUR-DIGIT CODES

We don't want you to get tunnel vision and to always aim directly for a keyway with your lock picking tools. You should also know by now that *most* bypass methods are faster and easier than lock picking; not 100% but a fair majority. This chapter is here to show you that there are also other options besides picking and bypassing, and this chapter on codes could be a book unto itself (and it will be...soon). This code primer should help you start solving some problems ASAP.

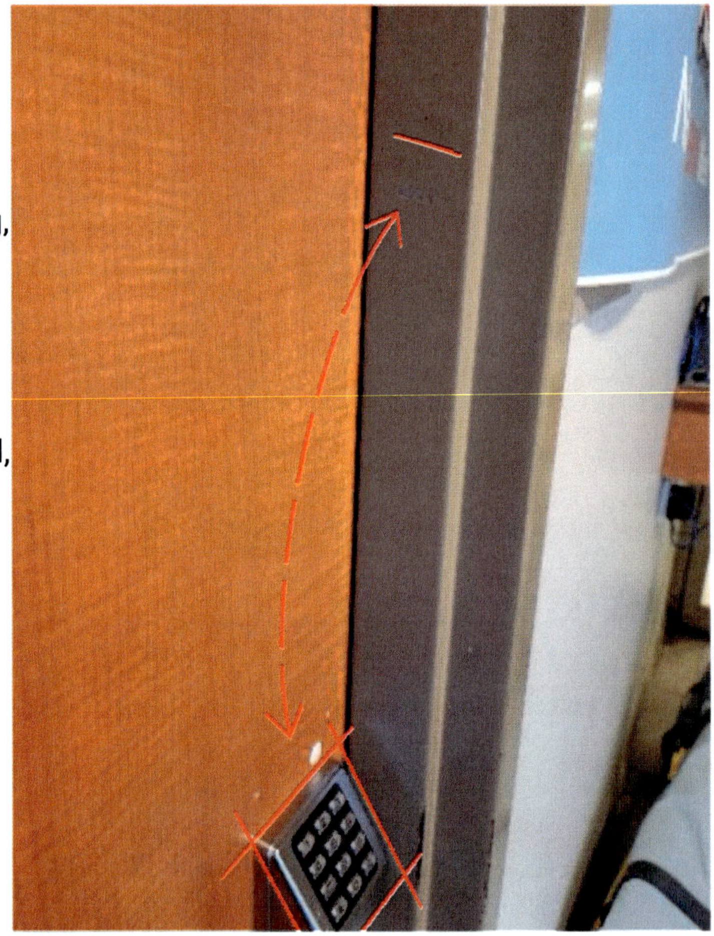

Isn't destructive entry so much faster than lockpicking though? Sometimes, even when forcible entry is authorized, there is some wiggle room that exists where a trained and experienced eye can take a look at a combination lock and can know in an instant exactly what the correct passcode for that lock is, or they may be able to narrow down the possible options quite a bit.

Sometimes it's worth saying "Oh! A push-button lock (or a keypad for a code)! Let me breathe for half a second and look up, down, left and right. Annnnnnnnd there's the code, written down nearby." (Two step rule, overlap)

Ca-click! Entry!

This is not just something reserved for the super-spies of the big screen. This is real life. Just like in the chapter "Locked In" we saw that the combination lock was left on 9-5-8-0, and the address was 9-5-9-1.

Other times, especially with organizations that have a lot of employees and a high turnover rate, and access that is needed at all hours of the day/night, passcodes get written down in pen and ink, often within arms length of the locked obstacle's keypad.

I've also walked through prison gates, law enforcement evidence room doors, secret federal agency warehouses, overseas high-security military bases, sensitive hospital trauma centers, and more, all of which have been "locked" behind combination locks that fit within all six of our Attack Vectors below. This chunk of our curriculum that we teach comes from almost two decades of a wide array of documented field data. It is the *rarest* of cases that combo lock codes are selected randomly and that the security of that code is controlled and protected.

Understanding how to look at the coded world differently can help you save time, save money, and save lives.

I don't know if the claim is true but it sounds true enough to share, that "The number one activity that results in Law Enforcement line-of-duty injuries is the act of climbing or jumping over fences." Whether true or not, my old-ass knees and back hurt every morning and while I can still jump and climb and fight and shoot, I'd still rather pick open a lock, or decode a combo and walk through a freshly unlocked pedestrian gate, than to climb and jump over a fence as long as there is no dire life and death emergency. Tactical Lock Picking, and this code breaking skill, can help save you from the "I thought it was just going to be a simple fence to climb and I ended up in this cast for six months" occurrence.

For your skill development within tactical lock picking and entry studies, if you already have lock picking tools and training on how to use them, but no understanding of code breaking, why would you buy yet another rake and tension wrench set? The option exists for you to rapidly expand your capability with a notebook that has literally every single possible code for millions of push button combo locks. As of now, this is the only one of its type in the industry that we know of. In all honesty, our "Brute Force Notebook" is one of our absolute favorite projects we've ever released.

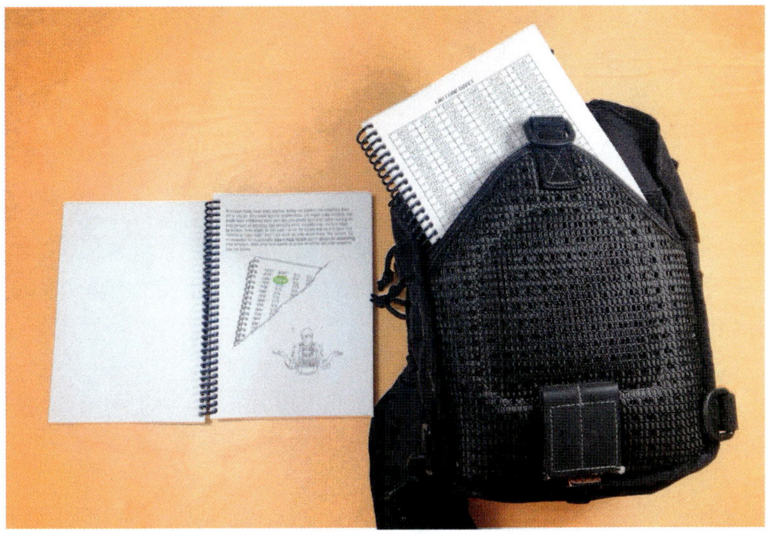

ATTACK VECTORS

What are some of the different avenues of which we can think about attacking four-digit combinations?

In our curriculum we have six different ways that we organize and group our Attack Vectors for when we decide to attack a locked obstacle by figuring out its four-digit passcode.

Socially Engineered Codes: We can attack the code of a lock by understanding that sometimes codes mean something to the locked obstacle or the user of that obstacle.

Manufacture Default Code (and Master Code): Attacking a lock by simply trying a known code that comes pre-installed on the specific make/model lock from the factory.

Evidence of Code: We can attack a code by using any type of existing evidence that helps us determine what a lock's code is.

Common Access Codes: We can attack the code of a lock by trying entries that are either a written number pattern, or that trace a pattern on a keypad.

Brute Force Attack: Attacking a code by...counting!

Random Ass Codes: Just the absolute worst for entry specialists to try to solve. A passcode and its associated lock that has been randomly selected and that doesn't fall into any of the previous Attack Vectors.

SOCIALLY ENGINEERED CODES (SEC)

These are codes that people have chosen for a reason. This is one of my favorite things to teach people "Things that you already know, that you didn't know that you know."

- Birthdays (YYYY, MMYY, MMDD etc.)
- Anniversaries
- Addresses
- Phone numbers
- Other dates

Sometimes people choose codes that are easy to remember so that they can tell others and then they, too, can remember it. "You must be new," says the hospital staff to the new ambulance crew member that just delivered a patient to the ER "the access code to the secure entrance here is 3-6-7-* (EMS*), should be easy for you to remember that!"

0-9-1-1, 4-3-5-7 (H-E-L-P), and other "themed" codes, and codes that are associated with letters, also fall under this Socially Engineered Codes (SEC) Attack Vector.

Pop quiz: If you were traveling to attend a presentation being held at the splendid "Colorado County Chamber of Commerce" building, and the digital keypad on the front of the locked building had the paint from the #2 digit rubbed almost all the way off, what might be your guess at what the four-digit entry code is for this theoretical keypad? If you're not old enough to remember the old-timey telephones that had letters associated with each of the numbers on your phone, the digit '2' used to represent the letters 'A', 'B', and 'C'. Hopefully you've started formulating a plan for how to gain (moral) entry into the CCCC building... that has the #2 digit worn from over usage...

If the code for this building *did* turn out to be "2-2-2-2", then that would mean something for the user or for the lock. It would be a code selected for a reason.

[Opposite: Imagine you were visiting this store on the right, established in 1998, last four digits of their telephone number is 7777, and their street address is 2525 Main St. Imagine, too, that when you bent downt to look at the key lock-box that was hanging on the door handle, the combination code on that lock box was currently showing "7-7-8-8". Even if you had permission to employ destructive entry tools to save a life inside, don't you think that there is in fact enough time to rotate those last two dials down one number each to try something that, as a well trained eye, is staring you right in the face?]

STORE
Est. 1998
555-555-7777
2525

MANUFACTURER DEFAULT CODE

Almost all mechanical combination locks (push a button or rotate a physical dial to input a code) come with a pre-installed passcode upon purchase. Sometimes, if you have to install the lock and set the first combo yourself, a manufacturer-recommended test code is usually listed in the installation manual which, after testing, you're *supposed* to change...

There are also a lot of digital locks (push the buttons on the electronic keypad and they usually beep beep beep) that have a factory-installed default code as well as the occasional master code.

> Defined: For *mechanical* combo locks, they almost always are delivered to the end user with a **"Manufacturer Default Code"** already installed before the lock leaves the factory. This MDC is usually written somewhere on the packaging or listed within the installation manual. Physical/mechanical combo locks almost never have a "master code" and almost all of them can only have one passcode installed at a time. Exceptions do exist.

> Defined: A **"master code"** for digital locks will open the locked obstacle no matter what the current passcode is set to. Think of a digital master code as a "digital back door". Exceptions do exist.

Manufacturers of mechanical and digital combination locks will often post installation manuals online where you can locate those codes. In my courses, I recommend following a specific order to figure out the most likely and most efficient entries for four-digit codes, as well as picking open and bypassing locks. If you don't have a bypass option or a tool on hand for a particular bypass, then this order of operations often (but not always) starts with trying to locate a manufacturer's default code.

One of the most important field-use lessons that we didn't want to leave you without:

> When you have determined that your current attack vector should be to try to attack the four-digit code, before you adjust any of the wheels on the combo lock, **take a picture** of the current code that is showing on the physical lock (**or write down the code**).

This documentation is so that if your lock says "7-7-7-6" and you immediately grab the lock and adjust the wheels to say "0-0-0-0" (a common MDC) you've lost a data point. If the zero's don't work, and if you can't remember what was showing previously, then that 7-7-7-6 is no longer available to your brain as a data start point if, perhaps, some type of social engineering piece of data for your target lock is very similar to "7-7-7-6". Don't throw away data.

Many MDC's are open source and able to be looked up online with a simple internet search from manufacturer's websites and in installation manuals. If those don't work, many internet message boards will have that information or you can otherwise acquire it.

On scene, if you suspect that a lock has a MDC or a master code, it might be worth doing some internet searching to determine what those default codes or master codes might be. As always, make smart and reasonable use of your time on scene in relation to your mission.

1234
Installation Manual
1234

EVIDENCE OF CODE

If you are reading this book in order (and you are under no requirement to), you'll remember that our TLP curriculum guideline "Two-Step Rule" is a handy way to look for a key or a code nearby the obstacle. That has a large overlap here when we are dealing with codes. Seeing either a conspicuous, or a non-conspicuous, four-digit code written or posted within sight of a lock could, in fact, be evidence of a code.

If you have four adjacent wheels on a combo lock, evidence of a code may also be that the user of that lock didn't alter all four of the wheels after locking it, so the wheels will be showing a partial code. It is common for people to only change one or two wheels of a four wheeled combo lock. This, too, can be partial code evidence.

The most impressive application of EOC (Evidence of a Code) is seeing a worn keypad and being able to instantly mathematically determine all possible entry codes based on that evidence. Being able to understand the math behind figuring out the most likely passcode for that lock is a Fourth Line Gear skill; it is information based, free to carry with you.

Here is an easy foundational crash course for a perfect setup to using the EOC Attack Vector by seeing evidence of a worn keypad, and the math that goes with it.

MATH CRASH COURSE:

We're looking at a digital keypad. There is a label printed and stuck above the keypad that assures us that the code is in fact *four* digits long. The evidence also clearly shows us that four separate input buttons are highly-trafficked and likely the only four buttons used in the code. Because of this, we're going to assume, for now, that none of the digits are used more than once each (no repeating digits). [Photo: right]

On the opposite page, we will show you a visual example of how to rearrange numbers (or items) in a format called a permutation, loosely defined as simply rearranging a group of items. Then we're going to give you a list of all of the possible codes using the four buttons that are rubbed off of the keypad.

If you have four billiards balls in front of you, laid out in a line, left-to-right... [Photo: opposite]

"Blue - Black - Purple - White"

How many different ways can you rearrange those four billiards balls into a new order?

The first ball can be a choice of any four, the second choice can only be your choice of the three remaining, the third choice will only leave you with two options left, the fourth is one option only. **4x3x2x1=24**

If you have to enter a four digit code, and only four digits are used, each digit is only used once, it is just like rearranging four billiards balls on a table...

Evidence of inputs used: #1, #2, #3, #4

How many different ways can you rearrange those four inputs into a passcode?

The first input can be any of the four digits, then any three, then any two, then one. The same math as our billiards balls example. **4x3x2x1=24**

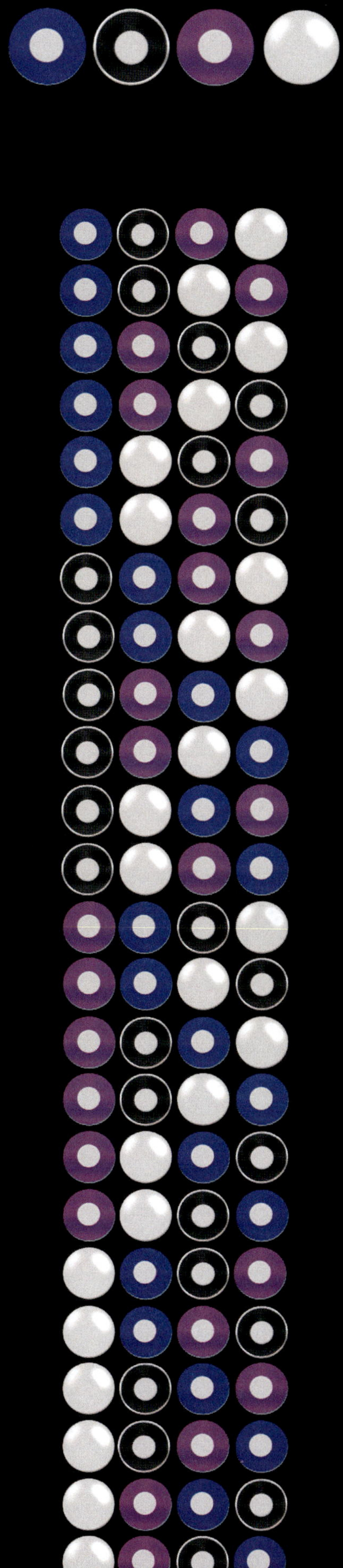

1234

Choose 4	Then 3	Then 2	Then 1
1	2	3	4
1	2	4	3
1	3	2	4
1	3	4	2
1	4	2	3
1	4	3	2
2	1	3	4
2	1	4	3
2	3	1	4
2	3	4	1
2	4	1	3
2	4	3	1
3	1	2	4
3	1	4	2
3	2	1	4
3	2	4	1
3	4	1	2
3	4	2	1
4	1	2	3
4	1	3	2
4	2	1	3
4	2	3	1
4	3	1	2
4	3	2	1

COMMON ACCESS CODES

If I can't look around or research a Socially Engineered Code, if I don't see Evidence Of Code, and if the Manufacturer Default Code doesn't work, then I usually move to Common Access Codes next. This step is usually just preceding any type of Brute Force Code attack, since a CAC attack doesn't take too long and it still allows us a high percentage likelihood of success with a minimal time and energy footprint.

Our curriculum groups our Common Access Codes into two types of formats. One group consists of numbers with a written or spoken pattern, the other group consists of a geographical pattern.

(COMMON WRITTEN CODES)

These are strings of numbers whose pattern becomes clear when you write them down or say them aloud. In other words, you will see or hear the pattern. These may not have a very clear shape or pattern on a keypad.

- 1-2-3-4 (Counting up)
- 4-3-2-1 (First four digits, counting down)
- 2-4-6-8 (Even numbers, counting up)
- 1-3-5-7 (Odd numbers, counting up)
- 6-7-8-9 (Last digits on the keypad)
- 7-8-9-0 (Last digits, including zero)
- 9-8-7-6 (Counting down)
- 5-6-7-8 ("Who do we appreciate?")

(COMMON GEOGRAPHIC CODES)

Think about tracing a "geographical" pattern with your finger on an actual key pad, like on the opposite page. Disregard the numbers for a moment, and just look at the shapes that your fingers would make.

- 1-3-7-9 (The four corners)
- 1-3-9-7 (Continued...)
- 1-7-3-9 (Continued...etc.)
- 2-5-8-0 (Straight down the middle)
- 1-4-7-* (Down the left side)
- 3-6-9-# (Down the right side)
- Other shapes of keypads may have other types of geographical patterns that pop up

As your experience grows you'll start to "have an ear" for when people drop their four-digit codes to one another in public. And with that new ear, you should keep a physical or digital notepad handy; even if you don't write down exactly whose lock it is, or where the lock is, you should be documenting as many codes as possible. Hear a code, write it down. Hear another, write it down. You'll start to see patterns emerge pretty quickly.

As I was engaged in the draft-writing process for the first edition of this book, I was walking down a hallway inside a public area of a large facility, and I heard two employees right in front of me talking loudly to one another. One said, "Amazing, my friend is such a child. He's damn near 40 years old and he just told me that for his bank accounts, and basically any four-digit code he has, his combination is **6-9-6-9**! Some people!" This is not made up.

The above is a real life case study that I heard with my real ears. I did not include it just because it's an adult themed reference. I included it because it's an extremely common four-digit code and it takes almost no time out of you day to dial it into a four digit combination lock, or keypad, that you may be struggling with. Even if it has half of a percent of working... if you make entry into 200 or more combination locks in your lifetime, then half of a percent means that 6-9-6-9 will work on at least one of those locks!

[Opposite: Examples of a code that shows a pretty easily recognizable pattern when written on paper, but that doesn't have a pattern on the keypad, and a code that doesn't show a clear pattern on paper, but that does have a clear pattern when typed out on its associated keypad.]

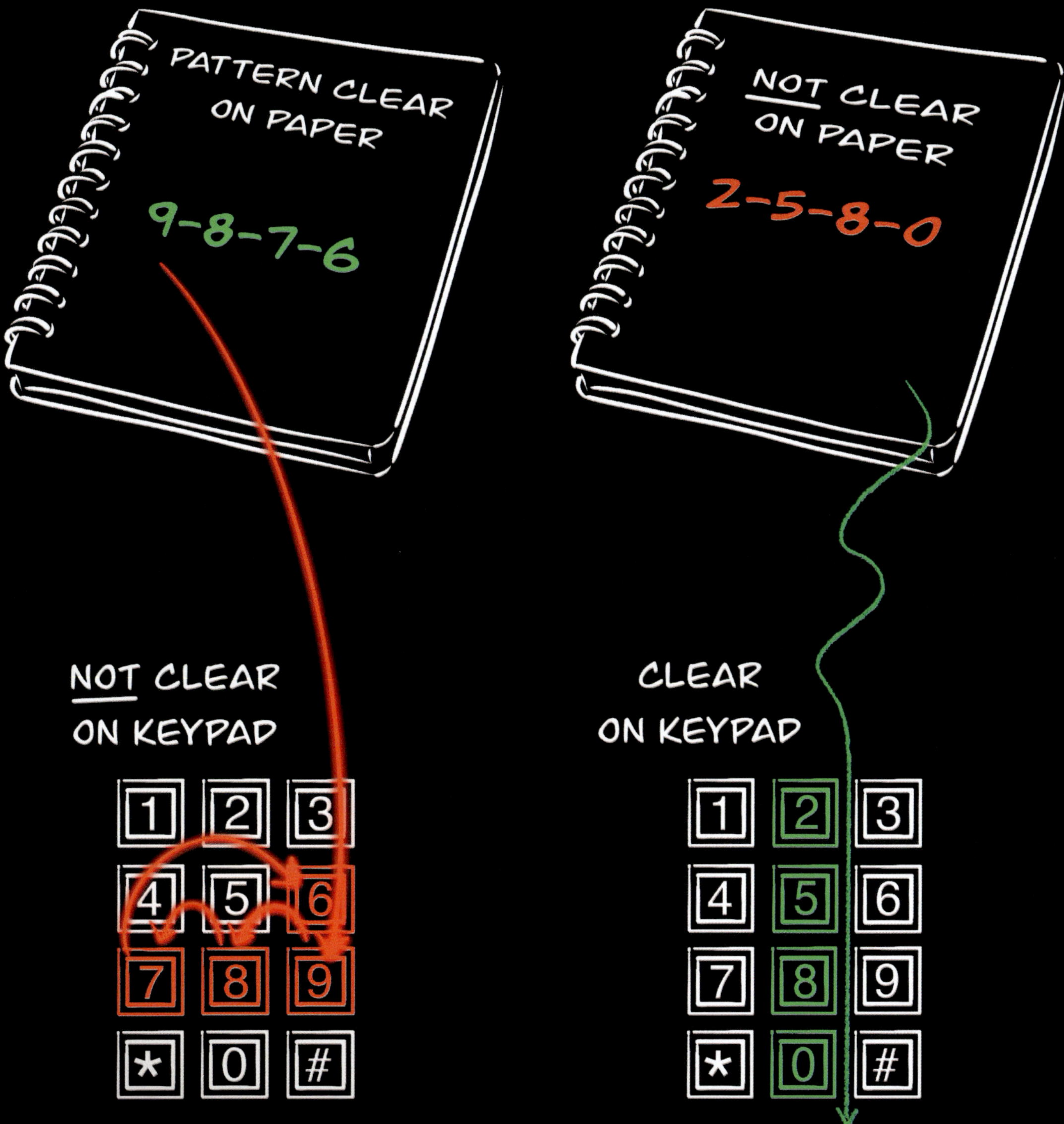
PATTERN CLEAR ON PAPER
9-8-7-6
NOT CLEAR ON KEYPAD
NOT CLEAR ON PAPER
2-5-8-0
CLEAR ON KEYPAD

BRUTE FORCE ATTACK

By definition, a brute force attack (not like sledgehammer force, but the force of math) is the act of trying every single possible passcode for a particular combination lock (or for computers, every possible arrangement of characters in a password). This definition comes with the defining characteristic of being 100% *effective*, but not always being 100% *efficient*.

Assuming no other variables are present (the lock isn't broken or rusted, you haven't missed a code due to operator error or programming error, or a flawed code sheet, etc.) then you can **100% guarantee** that the lock will open at some point with this method. A brute force attack could possibly be much faster than picking or bypassing, but it also could take much longer; **anywhere from the first entry or the last entry** in your brute force code attack could be the correct passcode that grants you entry.

Many variables go into weighing the tradeoffs of one Attack Vector vs another, such as different locks having different amounts of possible combinations. And remember, the value in Tactical Lock Picking is not in a selection of tools or techniques, but in how we make decisions and manage variables and tradeoffs.

For instance, picking a lock could be a high percentage of 80% likely to work, and the potential of an opening within a few seconds. That also means we have a 20% chance of never opening it via picking, or it taking a very long time. The brute force attack is an *almost* sure open, but at a much longer expected timeline.

Before starting a manual Brute Force Attack, always ask yourself if you've tried any of the other reasonable entry attempts first before you begin.

- Did you try to find an easier or unlocked way in? Did you try to bypass any locks? Did you look for a hidden key or code nearby?
- Did you try 1-2-3-4 or any other common/default codes first?
- Did you call for other resources to start making their way towards you while you work on guessing codes?

There are cases where a Brute Force Attack could be your best option, or your last and only option. Either way, it is still a force to be reckoned with. Being in this entry realm, there are very, very few tools that have a 100% effectiveness capability. This method, by *definition*, is one of them (again, assuming there are no hidden variables). Any time I can stack a tool in my gear loadout that is above the 90th percentile of effectiveness, I pay close attention when weighing if it belongs in my toolbox or not.

SOMETIMES THE ANSWER IS AS SIMPLE AS 0000, 0001, 0002, 0003, 0004... UP TO 10,000.

OTHER TIMES, YOU CAN USE YOUR DEPTH OF CRAFT TO TAKE WHAT APPEARS TO BE 10,000 POSSIBLE CODES, AND NARROW IT DOWN TO 210.

[Opposite: An example of the field usage of our "Brute Force Notebook" in action. A Brute Force attack is not always the best first vector of choice, but there are plenty of times that it can be an incredible resource.]

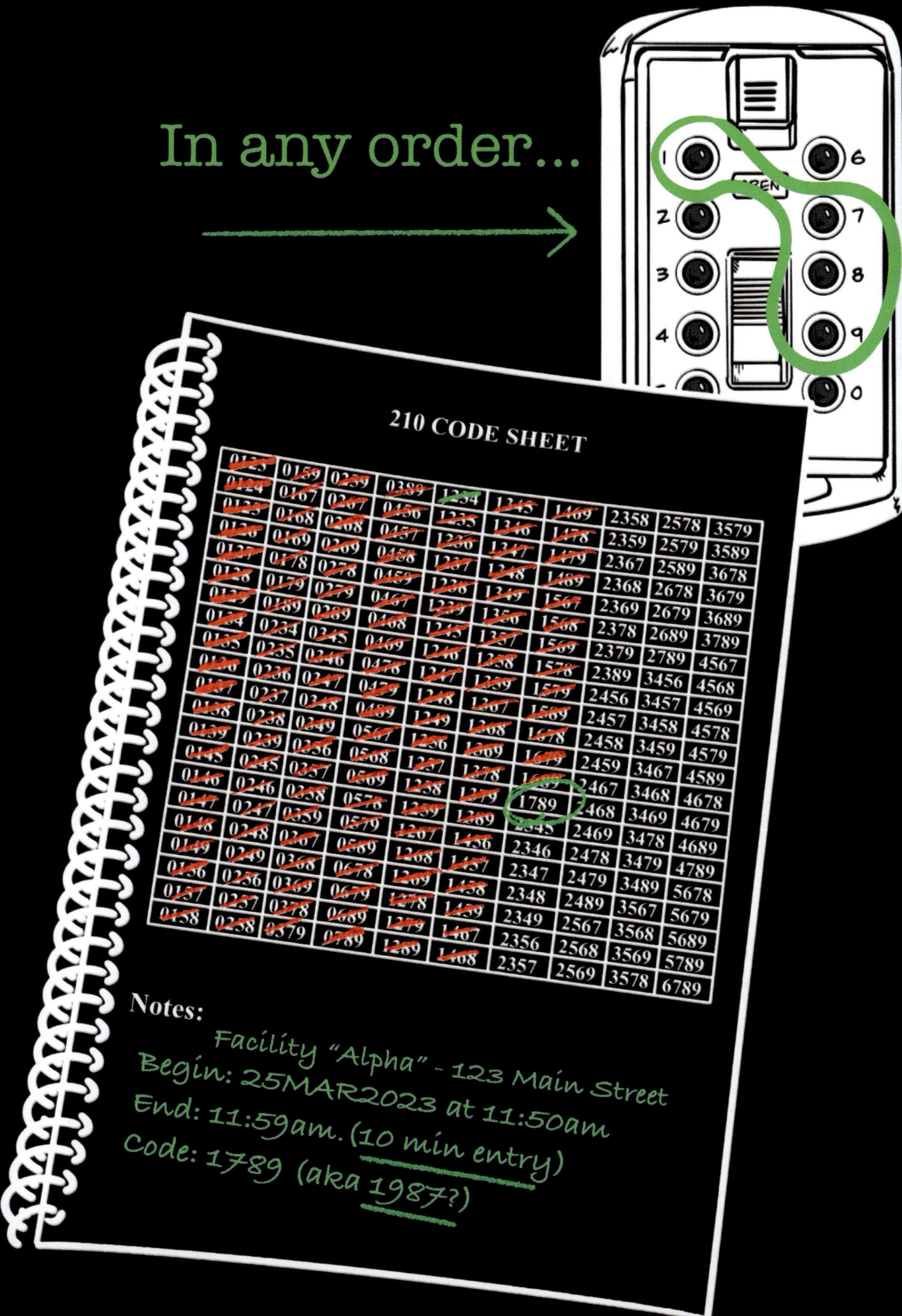
In any order...
OPEN
210 CODE SHEET
Notes:
Facility "Alpha" - 123 Main Street
Begin: 25MAR2023 at 11:50am
End: 11:59am. (10 min entry)
Code: 1789 (aka 1987?)

RANDOM ASS CODE

This is the area in this particular chapter where we'd usually tell you how to attack a code that fits the above subheading. In this case... not so much.

Random Ass Code defined. A code that intentionally, or even accidentally, does not follow a Common Access Code pattern, is not a Manufacturer's Default Code (or Master Code), is not a code that means a special thing to the lock or the user like a Socially Engineered Code, and where there is no Code Evidence that could be extracted to assist you in your code breaking investigation.

After trying other Attack Vectors and having no success, besides a Brute Force Attack, there's not much more we can recommend, as far as attacking the code. If you have a space that requires effective security, and it is protected by a code, ensure your code is random, and not easily discovered.

[If you are using our written works to assist you in teaching this skill to others (which we encourage, along with credit given to the source) feel free to call this Attack Vector the "Random *Assigned* Code" section ;)]

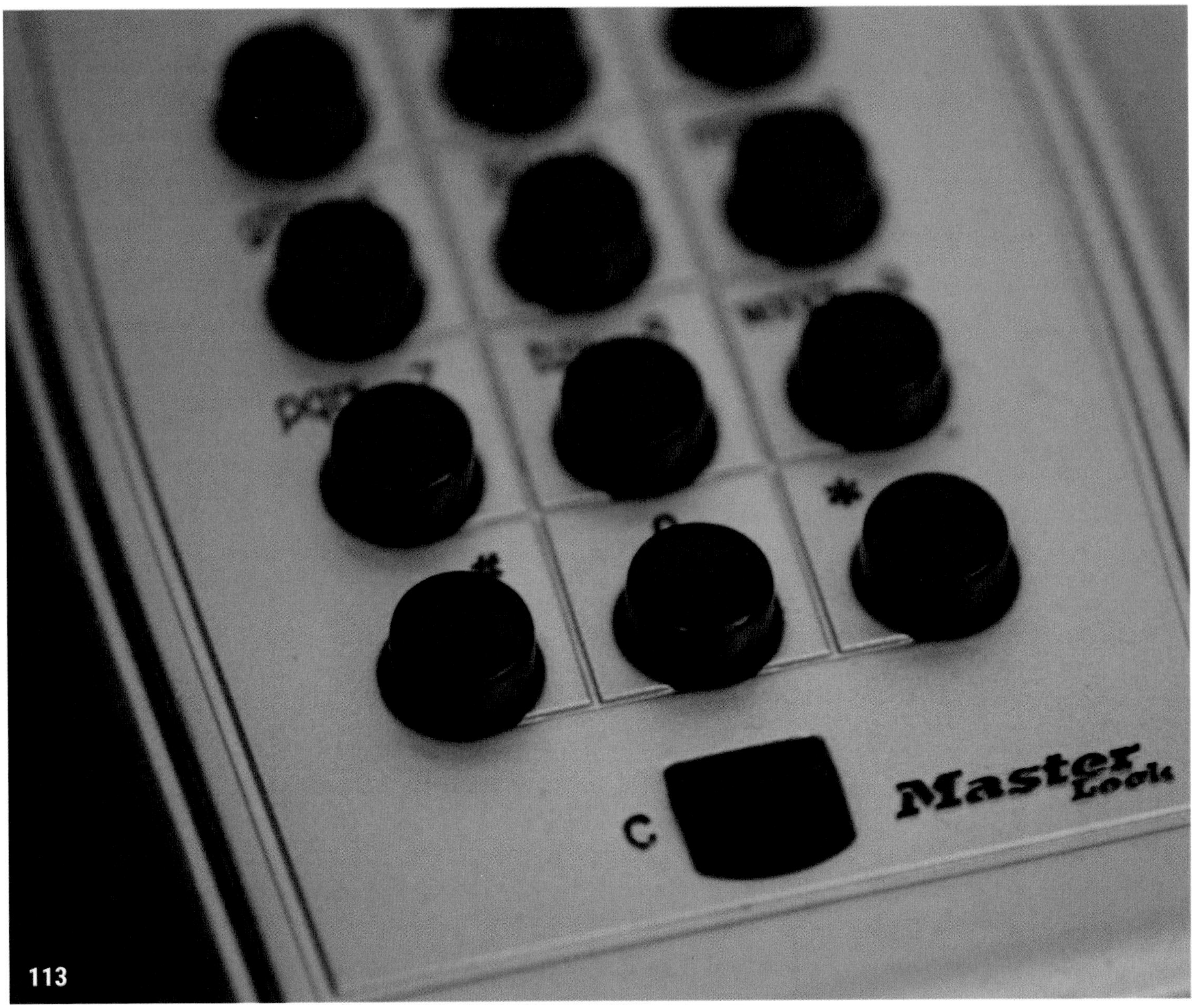

CONCLUSION

Every single practitioner of this skill set will have a different approach to how they train, what they research, which tools they carry, and how they make decisions on scene. For some of us that choose to nerd-out on a little bit of the math and the numbers, this four-digit code skill set can be an incredible resource. Even if you're not big on math, there are still ways that you can benefit from understanding how to do certain things like looking up a Manufacturer's Default Code within a lock's installation manual online.

None of this book, no skill in our courses, no block of curriculum, and no tool that we sell has ever come with a claim of "This one tool/technique/etc is the only answer!" When you see a four-digit code and you are on scene making emergency entry somewhere, we are most certainly NOT claiming that you should stop any other attempt in progress and that you should get out a notepad and a pen and to start doing math. You should always consider the factors that your specific environment presents to you, and you should always try to make the best use of your talent, your time, your energy, and your resources, in order to try to make **the most effective and efficient entry**, and a **reasonable** entry that is **appropriate for your current needs**.

There is an entire depth of skill waiting for you within the "just guess the code" concept, if you are interested in tumbling down that rabbit hole. And we're always here to help you do so.

[Photo opposite:]

If nothing else, that graphic on the opposite page should show you
the type of **depth and understanding of craft** that is *possible*
for those out there who which to become Covert Entry Specialists
or who whish to add Emergency Non-Destructive Entry to
their "useful human" toolbox.

The flow chart on the opposite page is **not** to be memorized!
It is a visualization to show you proof, within a concept, that we want you
making smart decisions so that you can make the most
effective and **efficient** entries possible.

We also don't believe in forcing practitioners in the field to
follow rigid guidelines just to "show their work". There is
no reason that you can't have an effective and efficient entry
while entirely breaking the traffic flow of the entry procedure
example on the opposing page.

If you just want to add a few basic approaches to tools and techniques,
we're here to help.

If you want to nerd out on the nuance of craft and the economics of different tradeoffs...
we'd love to train with you in-person and to learn from one another.

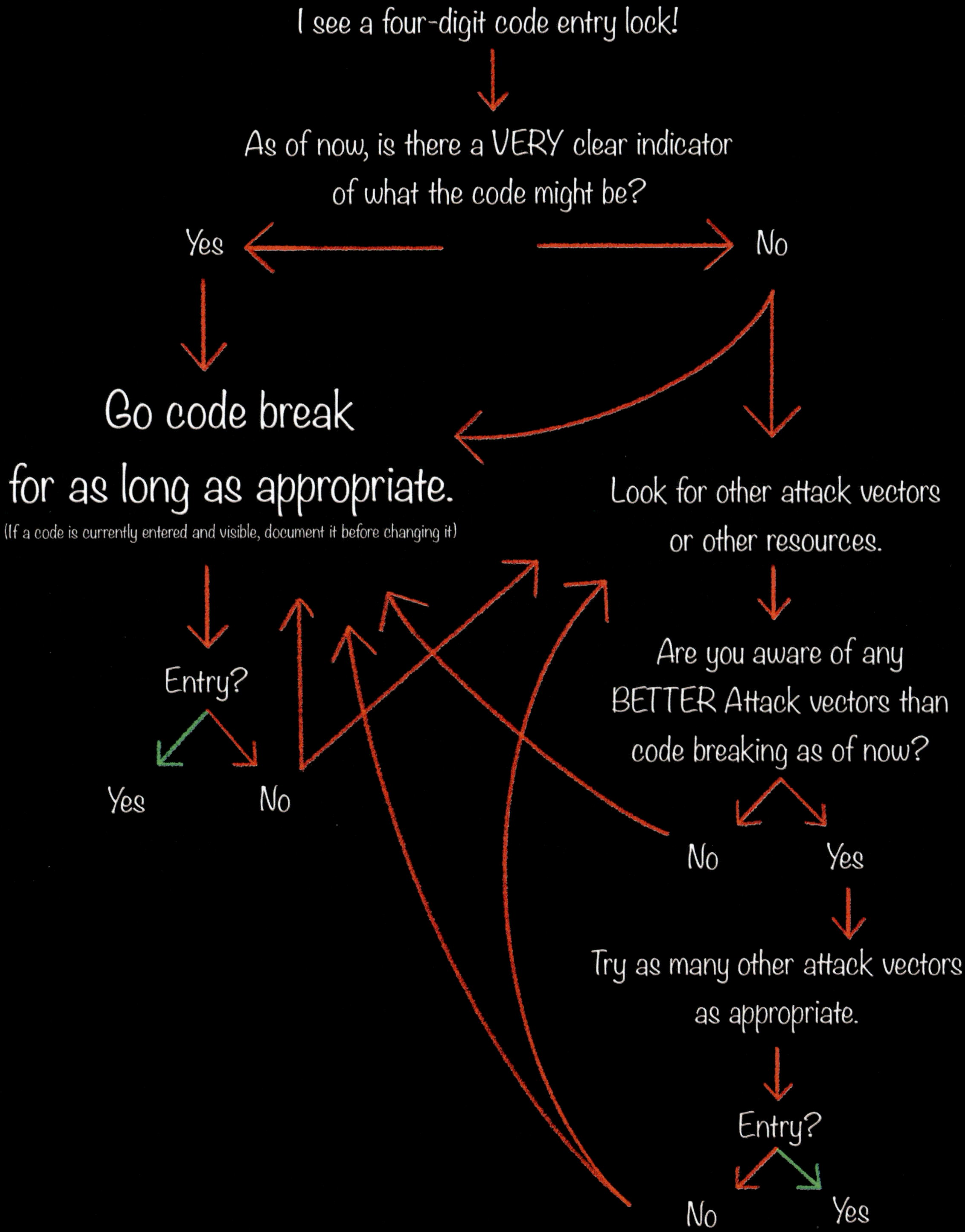

I see a four-digit code entry lock!
As of now, is there a VERY clear indicator of what the code might be?
Yes
No
Go code break for as long as appropriate.
(If a code is currently entered and visible, document it before changing it)
Entry?
Yes
No
Look for other attack vectors or other resources.
Are you aware of any BETTER Attack vectors than code breaking as of now?
No
Yes
Try as many other attack vectors as appropriate.
Entry?
No
Yes

CASE STUDY: CANCELED ENROUTE

When I was active in law enforcement, driving around in my patrol car, I would often receive calls from other deputies that they needed my lock picking capabilities to help them with something on one of their calls.

My standard approach used to be that I'd enjoy the drive to their location, I'd mentally skim through my Three Lines of Gear, and I'd happily start my entry decision making process once I got on scene... until that changed.

Case Study 1:

On a particularly lackluster case study, I was on patrol and got the call to assist with a lock entry. I drove across town, parked, got out with my Second Line Gear bag slung across my shoulder and walked up to the front door of a residence where someone inside needed law enforcement help.

The door was locked, I checked really well. But a few steps away I was able to slide open an unlocked window. This is still a win but, something was poking at me to make some updates to my standard response procedures.

Case Study 2:

Receiving another request for lock picking assistance, I attached to the other deputy's call and started driving. I decided to make better use of time for us, the responders, as well as for the victims that needed our help. I hailed the requesting deputy on the radio.

"Hey, I attached to your call and I'm 51 to you, about 10 minutes away. While I'm driving towards you I need you to try all the doorknobs around the house and to try sliding open all the windows."

It was about three to four minutes later that I heard the requesting deputy on the radio "Uhhhhh three-william-sixty-one? Yeah, you can go ahead and cancel," meaning that they had found an unlocked door or an open window.

Case Study 3:

Several times after that I had the ability to level up even one step more. My standard procedures changed. If I was off scene, and someone on scene needed my entry help, they were not just going to wait for my arrival, they were not just going to check to see if doors and windows were unlocked, they were going to be my eyes and ears, and my hands, until I arrived on scene.

- Send me pictures of each door, the whole door.

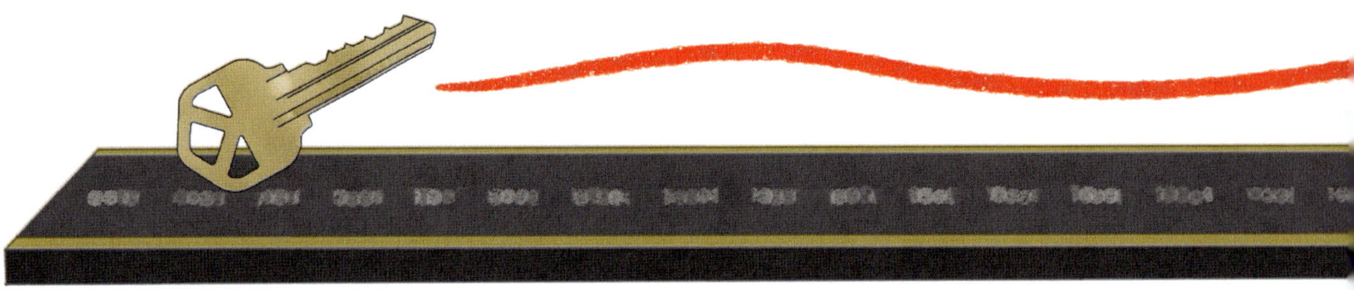

- Send me pictures of close ups of each keyway, deadbolt, knob, etc.
- Send me pictures of anything you find that is a four-digit padlock or door lock.

And if I noticed a possible Attack Vector where I could explain a verbal crash course to someone, teaching them how to use a particular makeshift tool or to apply a specific technique, I'd start having them attempt the entry themselves, before my arrival on scene.

CONCLUSION

TLP is a highly-transferable skill set.

- We can hand other people our tools.
- We can teach other people how to use our tools with short crash course lessons.
- We can tell people information, from a distance, and have them apply certain techniques for us.
- With things like our brute force code lists, we can tell someone, code by code, which entry codes they should try and in which order.

One of the most impressive aspects of non-destructive entry tools and techniques is that it almost never matters how much you can bench press. Non-destructive entry, covert entry, lock picking and bypass tools and techniques can be used by super spies, muscled up SWAT team operators, and soccer moms alike. The *nuance*, the *application of craft*, the *decision making process* is what sets apart all of the different practitioners of this type of skill set.

These case studies also highlight the fact that you might be on scene with your First Line Gear in your pocket when you realize that you need access to a locked obstacle in front of you, or you may be miles away in your car when you get called to a lockout with your Second Line Gear in the trunk of your vehicle. And what if your Third Line Gear, and other unique items not in your usual carry methods, don't live in your vehicle?

Sometimes decisions can be made as to which tools you pack ahead of time, or which tools you have sent for and brought to you on scene, or when you should put your entry attempt on hold to go retrieve more tools yourself.

Concepts about "how to make entry-based decisions" is what we want to highlight in this book and in our courses as well. It is all food for thought that will be dictated by your mission requirements, your skill level, your available support and resources, and more.

We want to create problem solvers, not just lock pickers.

CASE STUDY: POOL FENCE REDEMPTION

It was near midnight. Several home owners had called about someone breaking into cars in a neighborhood that Deputy T and I worked together. We had set up a perimeter and the suspect was on foot, still in the neighborhood. While we were patroling through the neighborhood, a homeowner in his boxer briefs stepped out of his front doorway and flagged us down. He said "He just jumped over that fence into the community pool!"

The pool area had a six-foot high fence all the way around and a locked gate in the front. This was not a simple waist-high fence we could have hopped over. When possible, avoiding the climbing of fences is preferred.

As a much younger man, when this case study took place, I disregarded the fence-climbing warning and I scrambled over the fence to go find the suspect. Turns out, the suspect had jumped the fence into the pool area but had jumped out, over the far side, just prior to our arrival. Wamp wammp.

The suspect was long gone so now we had some time. I thanked the Gods that I bounded the fence without-injury and decided not to test my luck again if there was no need to. Deputy T went back to the nearby homeowner to see if they had a pool key for the locked pool entrance gate. He did! Deputy T and his new friend arrived and tried the access card in the pool gate's card reader, but there was a time lock that denied access outside of the posted hours.

I decided to take a walk around the interior of the pool for one last check before having to climb this high fence again. I eventually found an inconspicuous maintenance gate on the far side of the pool. I tried to reach through the gate to activate the latch on the outside of the gate, but it was padlocked shut! I pressed my forehead up against the gate but couldn't identify the make/model of the padlock. So I pulled out my phone and reached through the gate to photograph the padlock. Looking at my phone screen, I was familiar with this make/model padlock and I'd picked open 100's of them before, and with the same lock pick set that I had in my wallet at the time.

Same Lock, Same Key guideline in effect. This was not a guarantee that I'd gain access but it was a really good starting point that steered my decision making. I felt confident, "Now's my chance," I thought to myself! So I called over Deputy T (from the chapter "Embarrassed") for a chance to redeem myself. When Deputy T arrived, standing on the outside of the pool fence, I told him in a serious tone "Okay, if this lock picking entry doesn't work this time, then I'll never ask you to watch me pick a lock again."

It was dark out, I was already nervous/embarrassed because I failed in front of Deputy T before, I literally could not see the padlock, and I had to reach through a gate to pick this padlock open by feel alone, without sight.

It was certainly a challenge. When the changing or adding of any variables occurs during your entry, it has the potential of exponentially increasing the difficulty of your entry. When I got my tension wrench and my rake tool into the keyway it certainly wasn't "easy" but there also wasn't a tantalizing, edge-of-your-seat story. I got it open rather quickly and was able to unlock the gate, walk through, and relock the gate upon my exit from the pool area.

Deputy T was supportive. He said that he believed in me the whole time and that my previous failed attempt hadn't tarnished my skill-set reputation. I was very thankful for his support.

[Opposite: a similar gate with a reach through scenario.]

CONCLUSION

- Tactical Lock Picking is not always just about gaining access into locked areas. Sometimes it's about helping you *leave* an area (as we also saw in the chapter "Locked In").

- As you progress in this skill, you'll start to recognize additional makes/models of locks naturally. You don't need to make flash cards or memorize lists, just go start picking and you'll start to say more and more "Oh, I recognize that make/model of padlock". And when you do recognize a lock, you may remember it being an easy lock to pick, or a tough lock. Either way, that information can help structure how you navigate your Target Assessment on scene and will help you choose which tools/techniques and Attack Vectors you choose to spend your time on and in what order.

- "Your couch is not the field." And your kitchen table or your desk is not the field. Practicing awkward angles and different tension wrench positions can really help you when field conditions force you to stray away from your standard favorite hands position.

- Not being able to see the lock that you are currently picking is a very common variable for people that use this skill-set in the field. Practice with your eyes closed sometimes, and practice with a visual blocker (with your eyes open) sometimes too. I've found that those are both two very different mental processes that are worth exploring during your practice sessions.

[Right: Not only did the scenario in this photo have a "reach through" unlocking problem to contend with, but it also had a padlock that had "shielding" installed. Shielding prevents easy access with lock picking tools and it also prevents a lot of destructive entry access as well for things like bolt cutters. Any time you get the chance, with permission, to practice a picking attack on a lock that has shielding installed around it, it's a fantastic way to test and to increase your field-problem solving skills! It was almost easier, for this scenario, for me to pick the lock with one hand; two hands ended up getting in the way!]

[Opposite: the inside view of the gate, you can see that the angle and workspace (and shielding) made this a very difficult position to pick a lock in.]

LEGAL VS MORAL

What do you need to know about laws and lock picking? First up, our company and our content never issues legal advice. Next, lock picking (as a hobby and a skill set to practice) is legal in *most* states in the U.S. and you can do some very easy homework to double check what that means for you. For the most part, the *intent* is the important part about carrying lock picking tools.

We recommend you *don't* rely on the first few blog posts that the big search engines push out. We recommend you *do* start with something easy like the the lists from either of the following sites below. Once you see the overview for your state in question you should then search the specific statute referenced for your state and verify the content of that statute on your specific state's government website.

https://www.toool.us/lockpicking-laws.php and

https://www.lockwiki.com/index.php/Legal_issues

Don't get overwhelmed, this example process is very easy and pretty quick. You can have a confident answer in less than five minutes; much better than reading a clickbait headline on a blog post and ending up less-than-informed.

Lastly, people also ask me a lot about travel, besides state-to-state. Specifically the TSA and flying. Long story short, as of the writing of the revision of this book (2023) TSA has not placed lock picking tools on the prohibited items list, and they have also specifically listed them by name as allowed items. But, if you're worried about travel via airplanes then the good news is that there is a long track record of lock picks moving freely about the cabin. No "sharpened" lock entry tools though.

Legal issues are always subject to change and this is not legal advice.

[Photos below: credit to https://www.toool.us/lockpicking-laws.php
Not to be considered legal advice! Do your own research!]

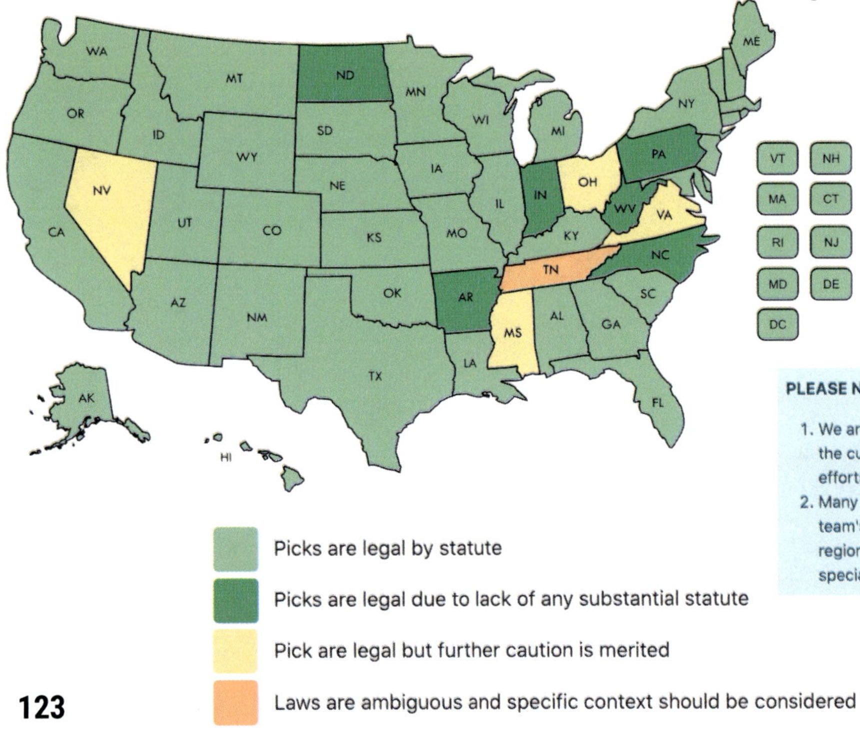

PLEASE NOTE the following two important facts:

1. We are not lawyers and this page is not offering legal advice. It simply represents the current laws as they are on the books at the time of our lawyers' research efforts.
2. Many municipalities have laws that may be different from state law. Also, our legal team's research focused on picks in general, not specific other tools. Some regions have separate laws concerning bump keys, auto tryout keys, and other specialized equipment.

EITHER, NEITHER, OR BOTH

Moreover, if one accepts the legitimacy of "laws" (political commands), one must also accept that drinking alcohol was perfectly moral one day but was immoral the next day - the day "prohibition" was enacted. Then, not many years later, it was immoral one day and moral the next - the day prohibition was repealed. Even the gods of most religions do not claim the power to constantly amend and revise their commandments, to regularly *change* what is right and wrong.

- Larken Rose, The Most Dangerous Superstition (2012)

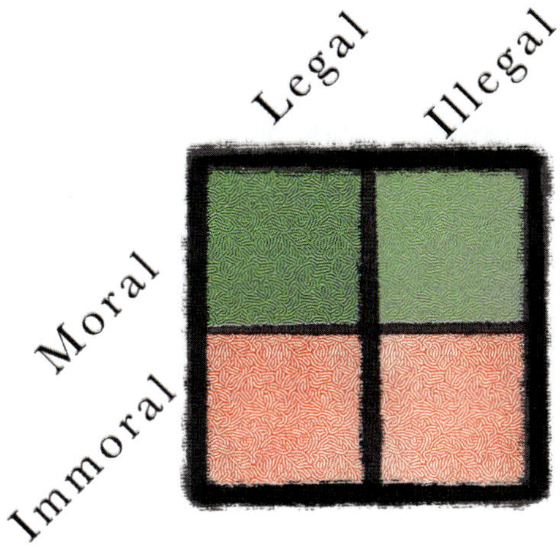

Legal and Moral

Surprise! An action can be either of these, neither of these, or both.

Explore this at your own pace.

PRANKS

Years ago, I attended a lock picking course that was restricted to students that were active law enforcement, and the course was taught by an instructor that was either active law enforcement, or prior law enforcement, I can't remember which. During the course he made the following statement, almost verbatim, "Now that you have these skills, go out there and go nuts! Go prank your friends, go break into your coworker's private spaces, and send us pictures and tell us about it! We love that!"

I've made it absolutely clear since the beginning, and continued a consistent stance for over the past decade that not only do I strongly oppose the violating of people's personal spaces for amusement, but that UTAC has been on record warning lock picking practitioners about the dangers of falling into the trap of "pulling pranks" against people using entry skills.

If you are known as the "lock-pick-prank guy", when someone notices that something is missing in their locked space, and you are nearby, even if you're innocent, you have painted a target on your own back. And, rightfully so.

Not to mention, imagine that of all people a law enforcement officer decides to prank someone by illegally and immorally breaking into their private space for their own amusement and they find evidence of a crime in the process. If you're not familiar with the court system, some fun concepts to look up would be "illegal search" and "fruits of the forbidden tree". Imagine finding something really bad, and not being able to prosecute someone because you only found evidence because of an illegal search. "Go and prank people, and send us pictures" what an interesting moral stance to take...

Don't prank people. It's exciting enough, fun enough, and rewarding enough being able to save the day with non-destructive entry skills that you shouldn't have to violate people's human rights just for a laugh. It's also fun enough and exciting enough through training or teaching others how to open locked obstacles by means other than a lock's proper key or code.

CASE STUDIES: VEHICLE ENTRIES

VEHICLE CASE STUDY: "SURE, YOU TRY"

A very long time ago, I kept my very first set of auto-jigglers in a plastic zip-bag in the glovebox of my car (we've come a long way since then).

I had just finished my first CrossFit workout and the cardio almost killed me. Someone at the gym had locked their keys in their car and I was happy to help. Except it turned out that I couldn't. I got the jigglers from my car, inserted them into the keyway, jiggled them around, and I was unable to close the deal, or rather *open* the deal.

A friend of mine was looking over my shoulder and asking questions about how the technique worked. While still trying to jiggle open the keyway, I explained the process. He asked me if he could try and I happily obliged. From my high horse, I assumed he didn't have as much experience as I did, but sure, I was glad to have a fresh set of hands give it a shot. I'm embarrassed to admit it but I thought or said something like "Well, you probably won't be able to open it, but good luck!" Ugh. Add another cringe moment to the historical record that I'll think about when I try to go to sleep sometime in the future.

My friend put the jigglers into the keyway, and I coached him to apply gentle turning pressure while moving the jiggler in and out of the keyway and up and down with a rocking motion while holding that turning pressure.

He paused occasionally, and I could see him trying to mentally feel his way through the lock. He said, "I think there's just one thing in there left for me to set." The lock turned over, and we all watched the inside of the vehicle door's lock indicator pop up and into the unlocked position.

[Above: Rule of thumb for using "auto-jigglers" for a vehicle keyway attack, the older and less expensive the car, the more likely a jiggler attack is to work. Newer and more expensive cars are using keyways that utilize different locking mechanisms, not "wafer locks" which jigglers are designed to open. Also, when appropriate, always try to grab a photo of the lock you've opened in the field, along with the tool that has opened it, recording your case studies (successes *and* failures) will help you grow in your skill set development.]

"SURE, YOU TRY" CONCLUSION

- Even though I was supportive of having someone else try the lock, I was less than optimistic about his chance of success. But from then on, my view of less-experienced pickers forever changed. I viewed them as on-scene resources and potential force multipliers.

- Curriculum change in progress. This case study took place way before the case study "Broken But Not Dying". By that case study, it was already standard TLP protocol to use the extra hands on scene and even to give them crash courses on tools and techniques when necessary.

- The act of teaching others how to perform certain techniques with certain tools is a skill worth developing. And staying humble in this skill set will pay you back in dividends.

[Below: A student in one of our Advanced Covert Entry courses using a vehicle unlock reach tool, inserted into a gap between the vehicle door and the vehicle frame. Used to push and pull buttons and to activate doorhandles. Reach tools are invaluable but they can't be carried in your pocket.]

VEHICLE CASE STUDY: "YEAH, BUT WHY"

The phone rang. "That's odd," I thought. TJ and I hung out on occasion, but we usually didn't call each other off-duty.

"I need your help," he told me, seemingly under pressure. "Being that I'm the new guy on SRT, it's my responsibility to drive the team to training and then drive them back to the station. And we're all hot as fuck and tired and sweaty, and yours truly locked the keys in the van. What do I do?" Yes! A puzzle!

While this wasn't a high-risk entry for me, it certainly was high-risk for TJ! At a minimum, he'd undergo some hazing and being made fun of by the team. So while TJ didn't have a lot of time to formulate a plan, I sure did.

"Let's get down to business, brother," I said. "You're training at the jail, right? That means there are probably a few similar fleet vehicles nearby. Tell me if you see any other county vehicles nearby that are the same make and model as the van you're driving."

He said "yes" but sounded slightly confused.

"Okay. Is there any way you can borrow the keys from some of the other fleet vehicles for a few minutes?" I asked.

[Above: The shapes, the "bitting patterns", for specific makes/models of keys are not infinite. If we throw out a large ballpark number of how many possible shapes there are for these vehicle keys, even with a large number, you're looking at somewhere around a 1 : 390,625 chance. And, how many vehicles that use this key blank are out there on the road today? Short answer, your car key *does* match someone else's car on the road. And if you add some "jiggling" to your technique, then that ratio goes *way* down.]

"Yes, but I can't drive the team home in another vehicle. All our shit is still in the vehicle I locked."

"I'm not gonna *steer* you wrong. Just answer the question. Is there any way you can borrow some of the other fleet vehicle keys?"

"Yes."

"Okay. Go get some keys, take them all back to the van you locked, then call me back."

Short pause...then another phone call: "Okay, I'm at the van with these other keys. Are you telling me that I should just try these other keys?"

Honestly, I hadn't even considered that yet. I was focused on guiding TJ through a specific technique but he wins a point for creativity because, believe it or not, sometimes your key will work as a 100 percent match with other vehicles. It happens.

I replied, "Oh yeah, of course. Try that first." Smooth. Real smooth, Pat. But alas, none of the other keys were a perfect match.

"Okay TJ. Next, pick any one of those keys and follow me step by step. Ready?'

"Ready."

"Put just the tip of the key into the keyway of your locked van."

Check.

"Turn the key towards the front of the van. So, counter-clockwise." This was a 50/50 shot. If it didn't work, we'd try clockwise next.

Check.

"Now keep a gentle turning pressure on the key while you push the key in and out, and move it up and down a little too. It's important to keep applying and hold that light turning pressure."

Check.

"Okay. Keep trying this technique with all the different keys you have. Try it a few times while turning the key to the left, then try it a few times turning it to the right. Call me back in a few minutes."

Check.

I kicked back on the couch at home and patiently waited for a response. If it was a negative entry, I could arrive at the training site with my tools in just about 15 minutes and still earn my $20 prize from TJ. If it was a positive entry I would earn my $20 without having to get off of the couch. That was my going rate at the time for off duty entry assists. Another perk to learning this skill set, besides helping others, is extra spare cash, or a free six-pack from time to time.

Between five and ten minutes later I got a text: "Dude! You're the man! I can't believe that worked. Saved my ass, big time!"

"YEAH, BUT WHY" CONCLUSION

I've used this "over the phone" technique for many different types of entries and so have several of my students. UTAC curriculum is designed not just for us to teach our students, but so that our students can teach others.

Tactical Lock Picking is not just a list of tools and techniques to practice. It's a primer to help students start to apply a systemized approach to solving problems. If a student shows up to one of our in-person courses and they have an expert-level skill in one or two techniques, we would not be impressed by that alone. If, however, a beginner student shows up and struggles with many techniques but in a scenario-test they apply a unique and unexpected solution to a locked obstacle, we celebrate that!

There was not a lot to learn from this case study besides "that worked!" The learning value is always greater from our failures.

VEHICLE CASE STUDY: "SET UP PERFECTLY"

Jordan, a long time friend of mine, one of my first ever Tactical Lock Picking students, was in the police academy several years ago. He got held up one day and ended up walking out to his car a little bit later than usual and the parking lot was a ghost town except for one car. There was another cadet standing near the car peeking in the window and two or three others were standing nearby seemingly trying to help with something.

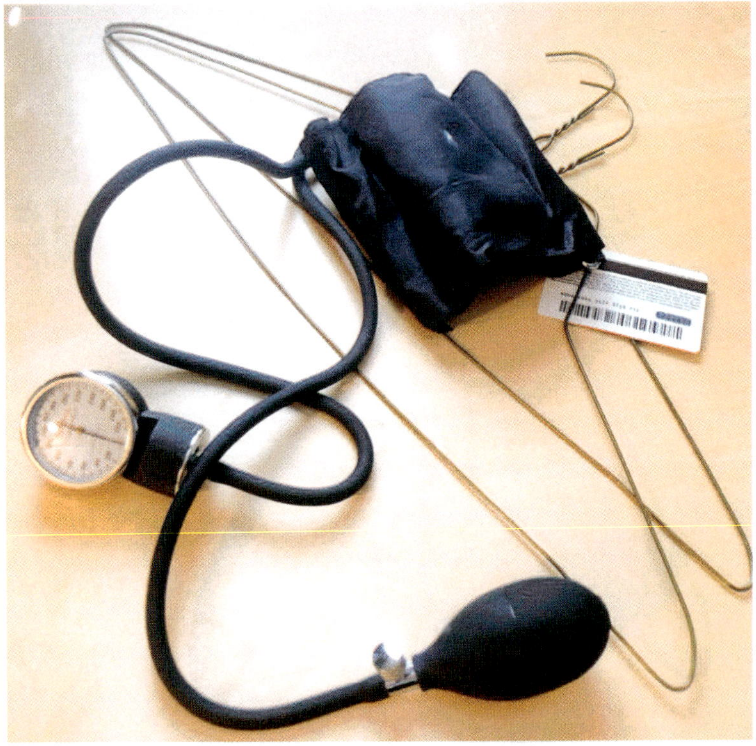

[Above: knowing how to create makeshift tools on the go is a fantastic piece of this craft that you can really get a lot of miles out of.]

He walked over to see what was going on. Turns out, the cadet had locked their keys in their car. It seemed that there wasn't much progress with the vehicle entry so Jordan started directing.

He turned to the onlooker wearing nursing school scrubs and said "Hey, I need you to go find a blood pressure cuff and bring it back here to the car please." He added to everyone else on scene "Go check your classrooms inside, and the closets, and your vehicles and bring me back all of the wire coat hangers that you can find."

The other students and cadets started returning with their items and it couldn't have been set up more perfectly!

Jordan took a credit card and folded the blood pressure cuff around it. This gave it more rigidity and allowed the BP cuff to be slid and pushed into the gap between the car door and the doorframe. The other students helped with their bare hands by prying the door away from the frame a little bit, making it even easier to get the BP cuff slid well into the gap between the door and the frame. [Photos below.]

With the BP cuff inserted, he pumped up the cuff which slowly spread the door away from the frame, little bit by little bit. Once a small gap of separation was established, Jordan took two of the wire coat hangers and straightened them both out, and then wound them together like the two sides of a strand of DNA. A single coat hanger is really flexible and doesn't make for a great tool to push or to pull. Winding two coat hangers together gives them a little more thickness and rigidity as well. [Photo: right.]

The keys were visible on the front seat of the car but the gap between the door and the frame wasn't large enough to loop the keychain and pull it out of the car (this was before keyfobs had an anti-lock feature if they are left in vehicles). So Jordan shifted targets, ignoring the keys on the seat, and got the coat hanger to push the vehicle unlock button.

"SET UP PERFECTLY" CONCLUSION

As with all of our case studies, the first two steps for every entry should be (1) Do we have the Moral Right for Entry and (2) what is our Speed requirement? Check, and check.

- Rule of thumb for vehicle entries: usually the older and less expensive the car is, the more likely you can use simple tools to gain entry.

- Beware: Most vehicle entry techniques are capable of causing damage. It doesn't happen all the time, but the possibility is always there, and there's a chance that it causes *expensive* damage. If the vehicle owner is present, make sure that they agree to the possibility that damage may occur.

- There are a lot of odd vehicle entry tools out there that claim to be magic. They are not. Be cautious about spending money on "magic bullet" vehicle entry devices. As with any tool, practice safely and smartly on as many vehicles as possible, with permission, before you have to rely on your skills and your tools in the field.

CURRICULUM: GATES AND CHAINS

Gates and chains are unique obstacles with unique solutions that have their own sets of patterns and approaches that fit them. The *art* of Tactical Lock Picking is not just about applying a tool and a technique. It's about problem solving. If you *only* solve a gate or a chain problem by trying to pick the lock then you're only scratching the surface of an otherwise useful process.

DAISY CHAINING

[Left: A recognizable padlock for even the beginner Tactical Lock Picking specialist. A "Master 175". Very easy to bypass, most of the time. But, should you sit down and start trying to bypass this very weathered and rusted lock just because you have an EZ Decoder in your pocket and you recognize the lock? Maybe yes, maybe no.]

A daisy chain, in our area of study, is when more than one lock is interacting with a chain (or other types of multi-lock accepting obstacles).

Sometimes, **more** locks equals **more** secure.

Sometimes **more** locks equals **less** secure.

This is not always immediately apparent to the untrained eye. Chains have a habit of hiding information when they are wrapped around posts, and they have a habit of making it unclear which padlock is, in fact, holding the locked obstacle locked. Sometimes padlocks are hung on chained locks and they are entirely useless. Other times, with multiple types of locks, if they all each grant complete access to the locked obstacle then that means you have multiple types of makes/models of locks that you can try to gain entry through. Always try to inspect the entire chain upon your initial assessment.

[Above: Always inspect the whole obstacle. Not just the whole deadbolt on a door. Not just the whole chain that is wrapped around the gate, but *the whole gate* as well.]

[Above: And a second lock, hidden behind the post. Very, very common with chains on gates. Now, you have the option to bypass or code break the Master 175, and you have the option to rake or pick open the keyway of this "WB" padlock (a very common padlock for public utilities).]

[Above: Annnnnd a third padlock! Yes! A version similar to the "American 1100" and the "American 700" padlocks. A lock that is above the beginner level of picking but a lock that is a pretty good benchmark in the locksport community as being a medium-grade. Completely accessible as a lock that you can pick open, it just requires some time and practice.]

[Right: For whatever reason (maybe if your tools or skills aren't able to unlock the padlocks or combo locks on the chain, or maybe your code breaking attempt might take too long, or maybe you don't have any "lock picking" entry tools. Sometimes, just regular tools might be the best bet to remove a locked obstacle. Crescent wrenches or socket wrenches might be all you need to take this obstacle apart. Start thinking about *specifically* which Attack Vectors are available to you and which Attack Vectors are the most appropriate for your current situation.]

For the daisy chain photo above, sometimes you'll have to access *both* locks on an obstacle in order to gain access. When you think of a door with a deadbolt and a knob lock, if both are locked then you'll have to solve both of those locked problems in order to get the door to swing open. This scenario, for chains (where you have to access both locks in order to access the obstacle) often happens on accident when the chain is already set up for a normal daisy chain with two or more locks connecting segments of chain, and a user accesses the chain and then re-locks the chain toghether *incorrectly* upon leaving.

Any time you get a chance to go inspect a chain that is holding a gate shut, go take a look and try to figure out which lock, and which *locks*, are keeping the obstacle locked. Even without picking any locks open, just the act of inspecting chains in the wild will start to give you usable experience and usable field data. Patterns will very quickly emerge and start to pop out at you.

[For the photo on the left, if you only open the metal colored keyed padlock, you won't have access to the locked obstacle yet.]

[For the photo on the right, if you only open the red four-digit padlock, you won't have access to the locked obstacle yet.]

HIDDEN LOCKS

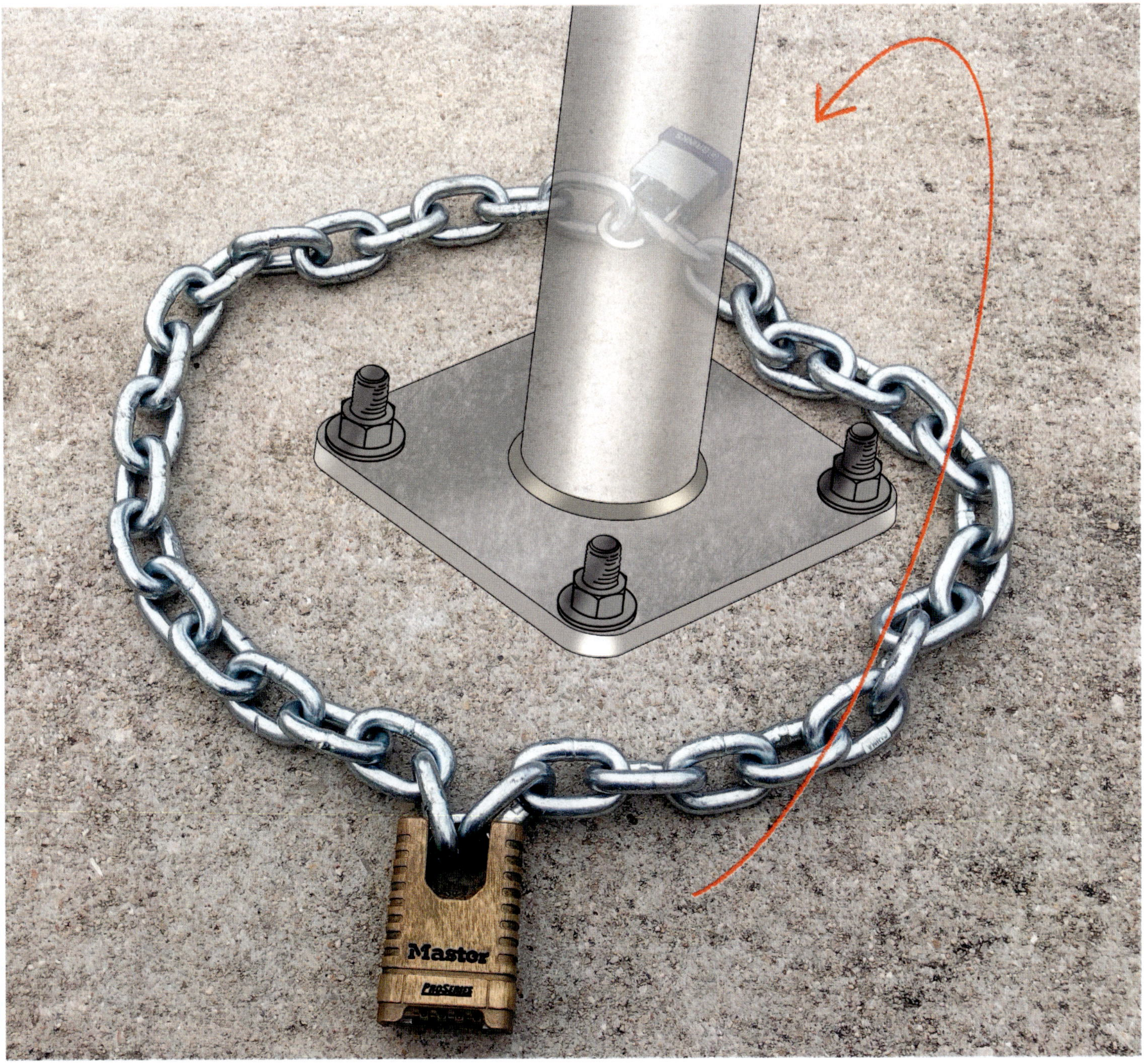

Very often, chains will have more than one padlock attached. A percentage of those times, the second padlock may be hidden from view. When you are assessing an obstacle with a chain on it, always try to put your hands on the chain and **inspect the entire chain**!

The above is a simulated layout. In real life, sometimes multiple wraps of chain can make an inspection of the whole chain a tedious process. Don't take for granted the total number of padlocks on a chain based only on the first padlock that you visually or physically identify.

FAUX-LOCKED

Faux-Locked Defined: A lock that appears locked, but actually isn't.

For both gates and chains alike, from a distance, they can sometimes appear locked until you make a much closer inspection. For certain types of locks, especially combo locks, their shackles may actually be closed but if the current correct combo is set then you may just have to push and pull the shacle for the lock to pop right open.

I hope that you are sick of hearing this, but always do a full and up close inspection of your locked obstacles (and beyond) when appropriate.

You dont' have to walk around a 100-acre facility before you start your entry attempts, but especially for chains and gates, put your hands on them and look up, down, left and right and don't get hyper focused on the first padlock you see. And don't get hyper focused just on the *locks themselves* either.

FURNITURE BYPASS

Defined: for any obstacle (gates, chains, doors, etc.) leaving the locked obstacled "locked" and otherwise manipulating the "furniture" that the lock is mounted to.

For the photo on the left, you could certainly find some way to electronically acces the gate opening system if you have that type of experience or skill set. You could unscrew the gate arm from its boxed mechanisms. Or, you could simply park, lift the ground-post out of the ground and drag it out of the way, then continue to drive your response vehicle to your emergncy. Expand your vision, don't get focused on only one type of locked access point, do a full inspection when able!

"REX" BYPASSES (REQUEST TO EXIT)

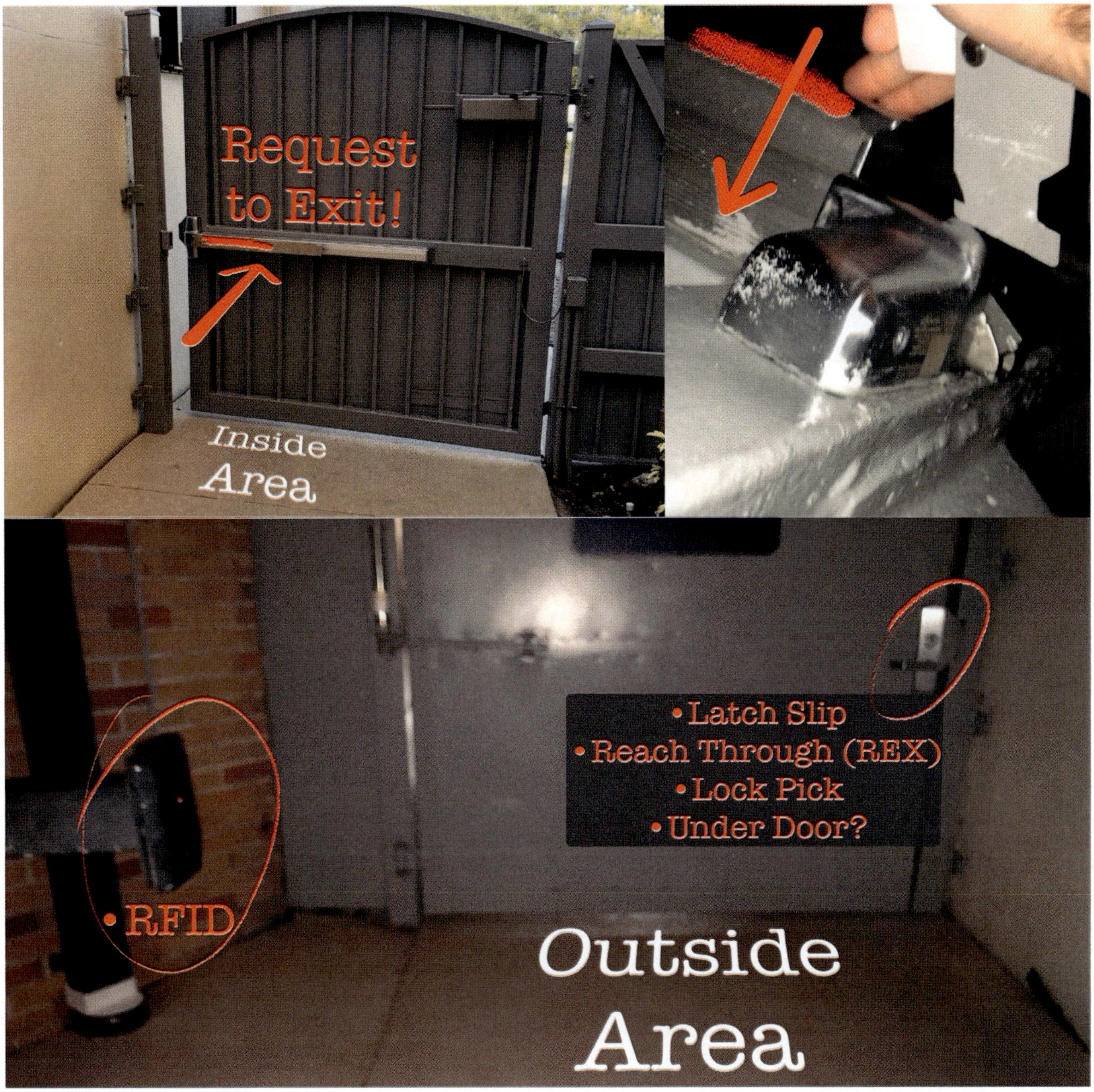

For the short study above, if you can reach through the gate from the outside to trick the locking mechanism into thinking that someone is "requesting to exit", that's often going to be one of your easiest entry methods. But don't forget that you still have other creative options depending on the tools you have available and the tools that you can makeshift on scene or have brought to you. It will also depend on your skill level, and it will depend on what Attack Vectors you are aware of. The appropriate Attack Vectors will also depend upon your skill level, and it will depend upon the context and the reason for entry.

If you're trying to get through this chest-high gate in order to stop an active shooter, go jump the gate (safely). If you need to wheel a gurney through in order to strap an unconscious patient to the gurney, maybe jump the gate first to render aid, and then push the bar to allow your other first responders access with the gurney.

There are too many "what if's" for us to cover here even on the simple gate problem above. The takeaways here are determining purpose, identifying options, and effectiveness and efficiency decisions for your environment.

GROUND LOOP DETECTORS

A VERSION OF "REX" BYPASSES

First thing to know about tripping ground loop detectors is that it is a possible Attack Vector...sometimes. This is the act of tricking an automatic vehicle exit sensor into thinking that a vehicle is exiting, and therefore opening the gate automatically (allowing someone to drive *in* from the outside).

The second thing you need to know about a ground loop REX entry is that it's *sometimes* possible.

Not every ground loop detector is an "automatic exit" sensor. Sometimes people in secure facilities can't just drive up to the gate but also have to "badge out" with a code or an RFID badge. So that already takes the social media comments, like "JuSt uSe a mAgNeT!" out of the game at least a little bit.

Additionally, the type of magnet that you use to trip these ground loop detectors is also important. Can't just roll some refrigerator magnets on the ground. The larger the magnet, the better, usually. So, again, this Attack Vector requires some real preparation and some real troubleshooting for the type of setup for the gate and the type of setup for the type of magnets that will and won't work.

You can also use *just* a metal shovel sometimes by scraping it along the ground on the ground loop detector area, but sometimes the shovel is not enough of a signature, or sometimes you have to get the placement of the shovel head *just right* in order for the detector to be tripped. Knock this technique down another percentage from the "100% magic" level. (NO technique in our curriculum is 100% effective. Always build in room for possible failures during your entry process and plan ahead to be able to work through them.)

And the timing is important too. How long will the gate stay open? Do you have a partner that can slip past the gate or climb over and activate the ground loop while you drive? Will you have to do this by yourself and to use a shovel with a magnet and to forcefully slide and send the shovel under a rolling fence and hope that it works? And, you should not be EDC-ing a huge magnet around. Even when stored properly it can still be dangerous to electronics that you own or dangerous to your physical health. While testing some of these methods on the job I have had these huge magnets smack to the side of my gun holster, thankfully not causing any injury or damage, but injury and damage are always possible with industrial grade magnets.

We aim to get our students to think and to solve problems, problems that almost always come with multiple variables. "Do I have a magnet with me? Where is it? What size is it? Is it big enough to work for that ground detector? Do I have a partner with me? Is this a solid impenetrable fence that I'll have to throw or roll or slide my magnet under? Or, will I be able to walk around the fence and then trip the sensor then run back to my car real quick? How long will the gate stay open? If I don't have a magnet, do I have something metal or can I find something metal to see if that works when I slide it over the ground loop detector? Is this even an automatic gate?" Variables. We couldn't *possibly* give our students or our clients "answers" that they can use when they come up against problems with variables, but we can certainly give them something better: a systemized approach.

Don't place your operational faith in tools and tactics that you haven't practiced with or experimented with. Always try to test your tools in training repetitions before having to rely upon them in emergencies. One of the nice things about this "tactical" skill set is that you don't have to wait for an actual gunfight to hone and perfect this skill. The data that you collect from a "peacetime" entry is the exact same data that you can use during a "combat" type entry.

[Photo opposite: A large commercial magnet successfully tripping an automatic vehicle exit gate's "ground loop detector." Stored in a *clean* sock for ease of removal from objects.]

VEHICLE POSITIONING

Referencing the example in the left half of the artwork below. If there is an emergency within a fenced-in and gated area, and you need response vehicles inside that fenced and gated area, it's likely that the first person that gets to the vehicle gate may not be your team's entry expert. It's also likely that the decision to smash the car through the gate, like we so often see in movies, is hardly ever an appropriate and reasonable tactic.

If this first person decides to take the option of parking in front of the narrow-gate and then decides to hoof it on foot to the location of the emergency within the fenced-in area, even if the second car on scene *is* the entry expert or otherwise does get the gate open, until that first vehicle is moved nobody else may be able to drive into the site because of the choke point and the first blocking car.

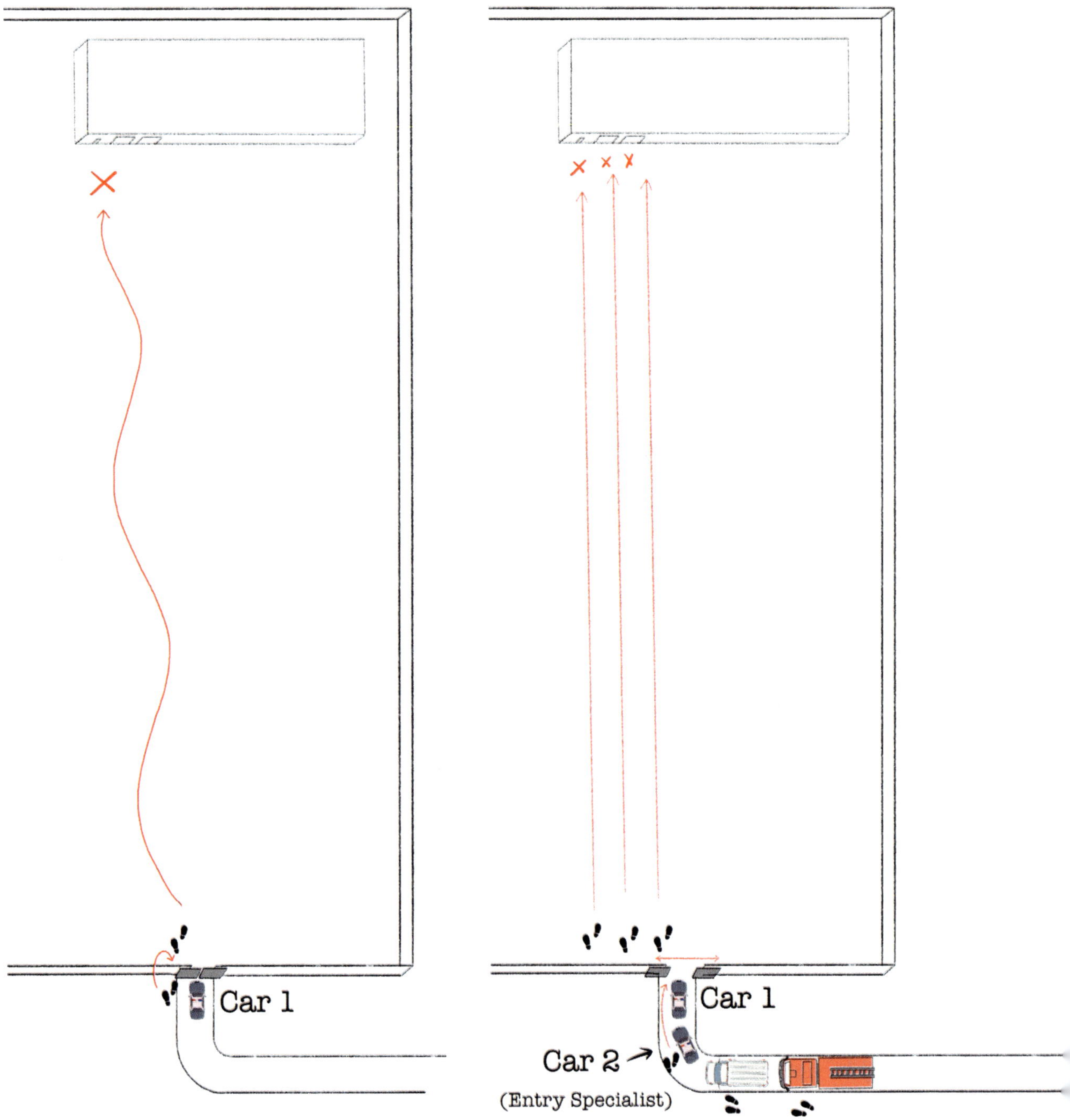

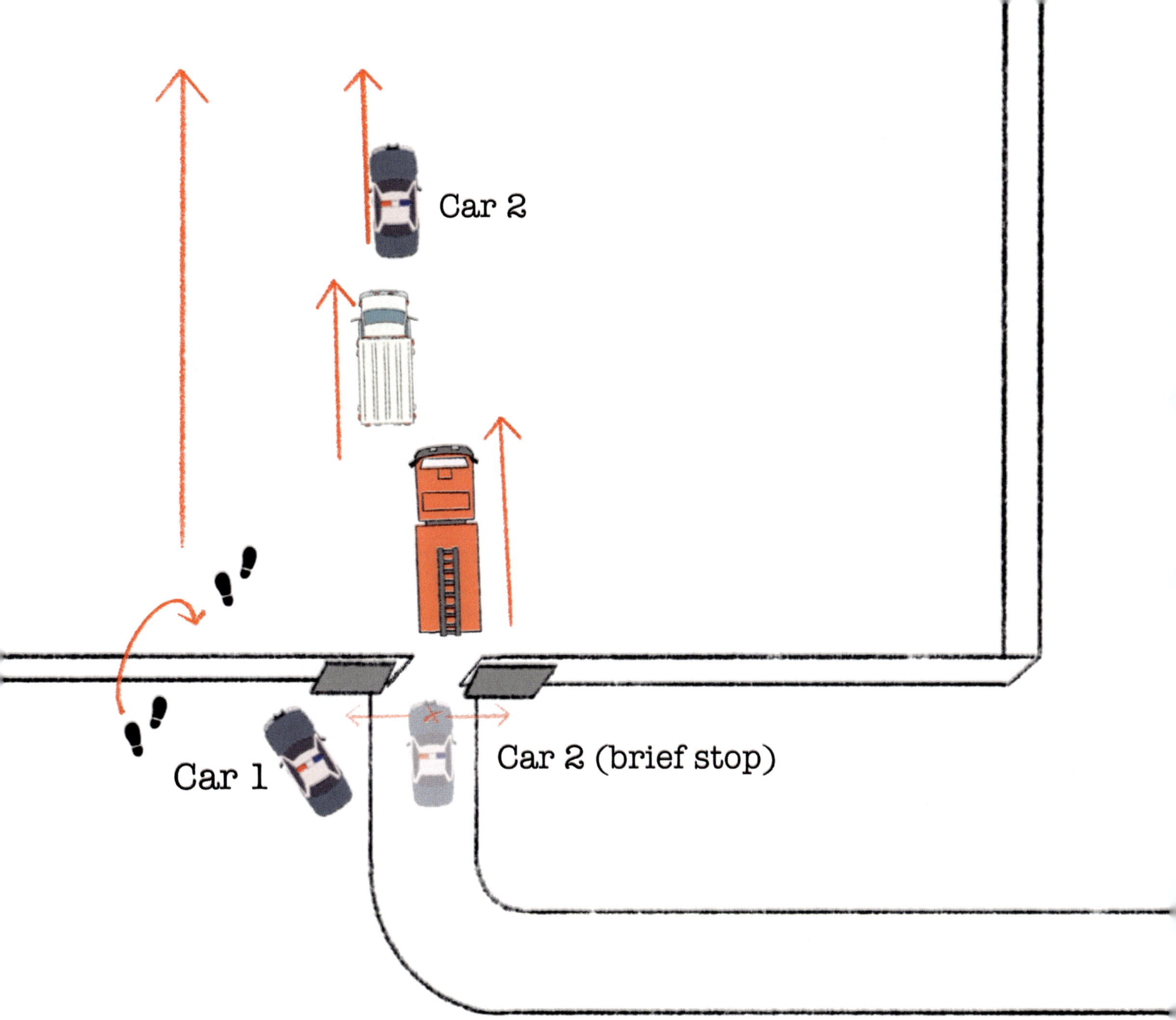

Here is where someone on a team that is merely “the one that knows how to use lock picking tools” may miss a chance to really support the team to the fullest, and where someone who “understands their depth of craft and understands a different way of operationally thinking through locked obstacles” can really shine.

In the artwork above, Car 2 (the entry specialist) hears Car 1 say on the radio “The gate is shut/locked” and responds “Car 1, this is Car 2. Make sure to park to the side and don’t block the gate. When I arrive shortly I’ll try to get that gate open for the rest of us. Go hop the fence and get in there and go start providing aid.” Even if Car 1 turns out to be the one to provide timely CPR to the dying victim, and keeps the victim alive, the entry specialist still gets to credit themselves with a very, very professional “assist” credit for that operation for getting the EMS vehicle up close to where the victim needs transported from.

A very simple action like reminding another unit where they might want to park so that different players can maximize their impact on the mission, that’s a huge skill set, and it has almost nothing to do with lock picking.

Go - think - differently. And if you want to really sharpen those skills and further develop those decision making nuances, we’d love to have you attend one of our courses or to have us come to your team/unit/agency to provide our training to you.

SKILL: RESTRAINT ESCAPES

READ FIRST

We're not telling you how to handcuff yourself and we're not telling you how to go about practicing breaking yourself out of restraints. You're going to have to develop and implement your own training plan since this skill could be dangerous if not practiced correctly. What we *did* want to show you, was that these techniques exist, they're real, and they can be accessed by people with very little investment in time/energy/resources, but just like in lock picking, these methods all require some in depth troubleshooting from time to time because none of these techniques are 100% effective, 100% of the time.

We also want you to know that "how to do the technique" will only get you as far as the first occurrence of struggle where you'll have to pivot your approach so that you can apply troubleshooting methods. Still, with restraints too, the value is in all of the failures you'll experience along the way.

We do offer a curriculum for a full 8+ hour day just on restraint escape techniques (and better ways to apply restraints), but we only scratch the surface in the "restraint escape" block of our foundational Tactical Lock Picking in-person course, as well as our self-learning video series that we've published "Tactical Lock Picking 101".

The best value that we can give you as part of this foundation of Tactical Lock Picking is to cover the most important parts of the most important tools and techniques. This is so that we have the time and the space to give you the information that really sets us apart, which is all of the ins and outs of decision making that will help you utilize this skill set in the field better than anyone who just "knows how to pick a lock" and then decides to apply that skill in a tactical role right out of the gate with no other depth or understanding of craft.

If you are going to practice this skill set yourself, here are some really important safety considerations that you should consider adding to your training procedures:

- If possible, always have a spotter.
- Always have professional restraint escape tools tucked into a pocket that is reachable while restrained.
- Don't practice in a location where if you drop your restraint escape tool you won't be able to retrieve it.
- There is a time for horseplay and being rough with students that are in restraints, and even for some advanced misery for the restrained student when practicing this craft, but that time is never to come without clear prior warning from all parties and clear informed consent before any rough hands-on training begins.
- Don't ever use a knife to try to cut someone out of flex cuffs, always, always try to find other more professional tools besides a knife.
- Start small, start slow, and gain a lot of easy successes before challenging yourself with more difficult restraint variables to overcome.

And remember, none of this is magic. We're just getting started and showing you that "these things are sometimes possible."

HANDCUFFS

Here's your crash course of information on how to remove handcuffs without a proper handcuff key:

- The handcuff design hasn't changed in any meaningful way for about 100 years.
- It's a simple ratchet system where all you have to do is fit *some type* of object into the handcuff body in order to lift up a spring loaded set of ratchet teeth.
- If the handcuff is double locked then the additional step of deactivating the double lock would require getting *any* type of tool to fit into the body or keyway of the handcuff in order to push a locking bar back a fraction of an inch.

In the U.S., handcuffs are not a restricted item. There are still physical stores that sell "security supplies" where anyone with cash or a credit card can walk in off of the street and buy police issue handcuffs and walk out without having to answer a single question by store staff or the cashier. We talk a lot less lately about politics/government but our general stance is that controlling items and controlling information often ends up victimizing peaceful people and doesn't provide any real guarantee of safety; here, we merely want to highlight that handcuffs are very widely available.

The fact that professional grade handcuffs are not restricted is just *another* reason to add to the argument that non-military, non-police individuals have every right to understand these lock devices and to understand how to unlock them, should they find themselves being illegally restrained.

If you believe that the general public should be forcibly prohibited from learning restraint escapes, then what you also MUST believe is that if one of your family members is kidnapped and illegally restrained then "it is much better that my family member stays kidnapped and exists in a condition where their escape is less likely - than to have a world in which civilians are allowed to learn how to hack a 100 year old simple lever lock."

The diagrams on the opposite page show the internal mechanisms of one of the most common standard handcuffs in use in the U.S. today. Using whatever makeshift tool you decide to use, and whichever technique you decide to apply with said tool, the *basic* application is very simple.

The real "ability to break out of handcuffs" is not free. It does not happen automatically after watching one how-to video on the internet.

Everything exists within a spectrum. This "ability", just like the ability to achieve a successful Tactical Lock Picking entry in the field, is only earned with experience. This ability comes with time and energy spent on research, information, practice, and the ability to smartly troubleshoot your tool/technique in accordance with whatever environmental variables you find yourself up against.

This is yet another spot-on example of how we try to frame all of our content, that tools and techniques always exist with a "percentage of liklihood of success" and that buying and carrying the best tools is only a small part of the game for increasing your liklihood of making entry through a locked obstacle or out of a locked restraint. How to open them depends upon the ability to properly troubleshoot.

Keyway
Pick

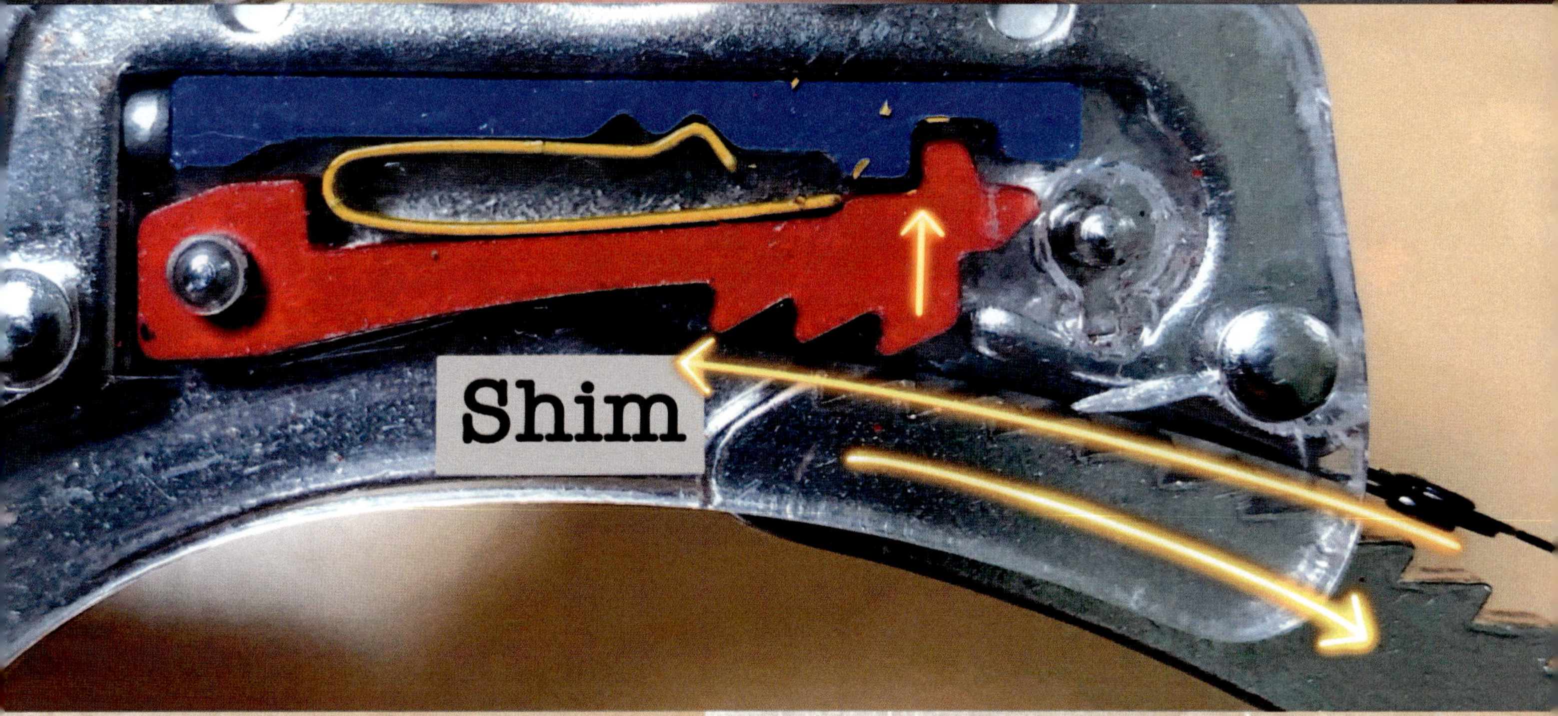
Shim

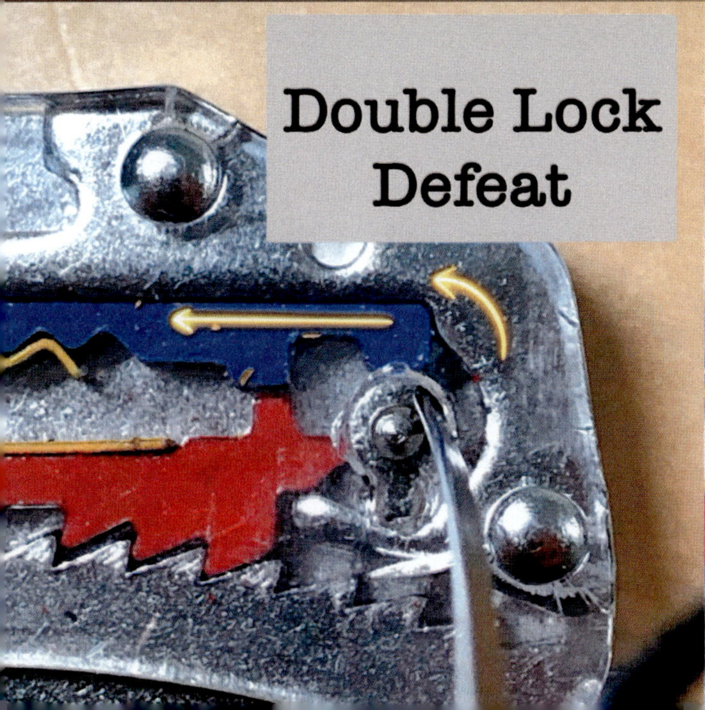
Double Lock
Defeat

FLEX-CUFFS

When I was active in the military and in law enforcement, I used to hold training courses for my teams and teach them restraint escape techniques for two reasons: (1) one was so that they knew how to escape their restraints if their own restraints ended up being used against them (2) and two was so that they had a better understanding of how to prevent detainees from escaping custody. When I showed my fellow officers a simple handcuff escape with a bobby pin, most of them had never seen the act done in person (let alone trying it themselves), but surprisingly when I pulled out a pair of flex cuffs their eyes often got even wider and I frequently heard statements like "Yeah, you picked open handcuffs, but can you break out of ::gasp:: flex-cuffs too?!" Yes. And so can you, and you won't have to look very hard on the internet to find *children* who are masters of these "secret" techniques.

Security is an illusion.

For the most part, flex-cuffs are just strips of plastic. There is no mystery, no mystique. The human mind is a strange thing. Let's change your perspective from "How on Earth would I break out of these big scary flex-cuffs" and let's switch over to a new approach of "How do I manipulate these pieces of plastic?"

You can **burn** plastic to melt it away. Friction is one way to generate heat without something like a hot piece of metal or a fire.

You can still interact with the locking "teeth" within most flex-cuffs (not all) by inserting a thin **shim** into the locking block area.

You can often perform a kinetic **break** by quickly applying an overload of force to the flex-cuffs (particularly in the area of the locking block and the teeth).

You can **cut** plastic strips if you have any type of *scissors*, although with cheap scissors it's often a difficult task.

You are only limited by your imagination. There are lots of ways to escape flex-cuffs and all other restraint types. But there is a big disclaimer when it comes to cutting: I would never recommend that someone use a knife to cut their way out of flex-cuffs (or to cut someone else out of flex-cuffs) unless it was a literal life and death situation. The way that a single sharpened blade interacts with most flex-cuff plastic is not usually a recipe for a safe or successful escape.

With flex-cuffs, and even othe restraints too, you can often **slip** the restraint off over your hand if they're not overly tightened.

As far as the "art form" of restraint escapes is concerned, this is a *very different mindset* from emergency non-destructive entry. The first goal is to never find yourself in restraints in the first place. And then there are manipulation tactics to hopefully fool your kidnappers into not applying the restraints correctly and to not allow them to find your hidden tools that you may be carrying. Then there is the timing of when you'll start your escape attempt. It's just a rabbit trail of questions that are certainly worth considering, and worth training if you so choose, but we just wanted to show you here that these things are possible.

Be smart. Be safe. Think differently. And if you want to practice restraint escapes then do so in a responsible manner and take responsibility for your own actions and your own training plan.

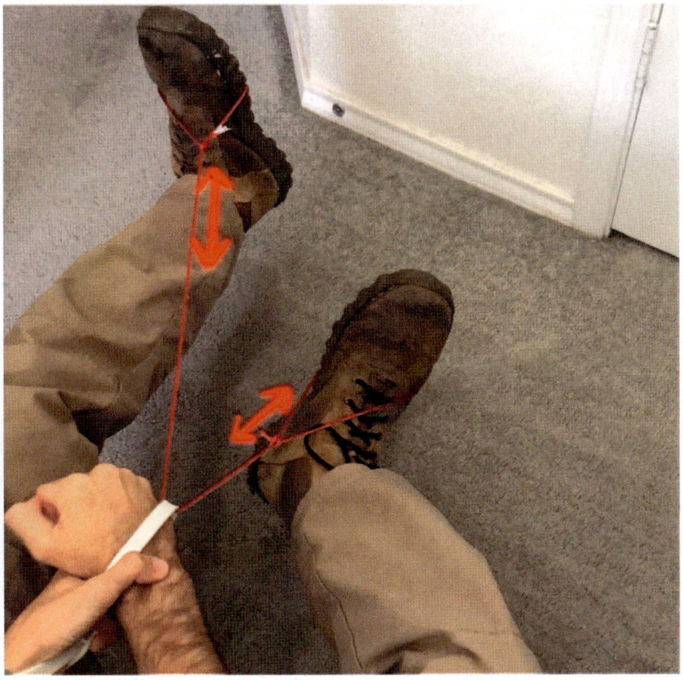

[Above: A perfect setup in a perfect world for a flex-cuff friction saw technique.]

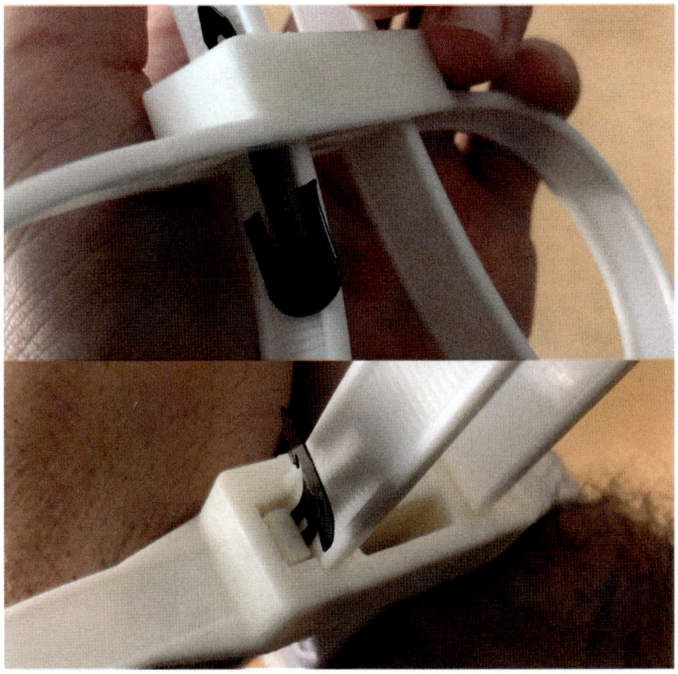

[Above: An example of a shim technique from the inside and outside of a flex-cuff.]

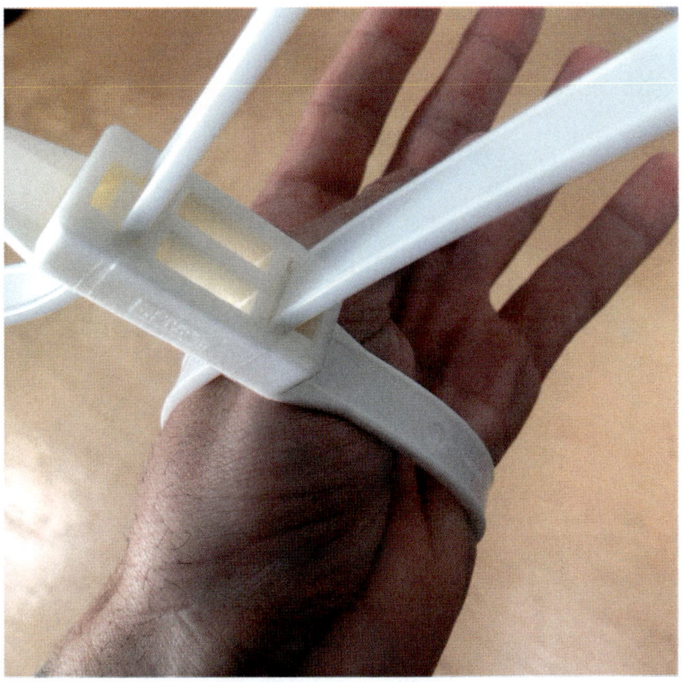

[Above: With many types of restraints, unless they are overly tightened, simply slipping the cuff over your hand is often an option.]

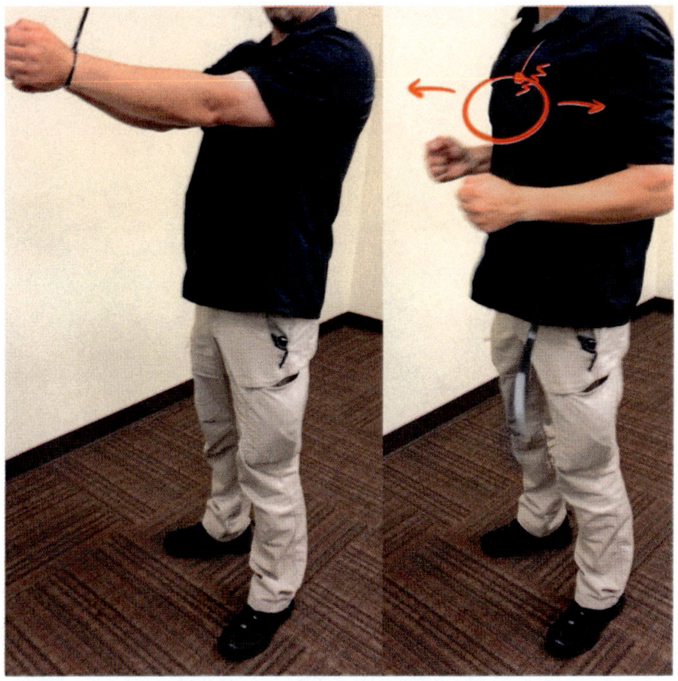

[Above: An example of a friction break with a small lightweight zip-tie. I would not attempt this with the larger bulkier flex-cuffs unless it were life and death.]

DUCT TAPE

The primary preferred technique that we teach to students in our course for removing duct tape is a sharp kinetic motion to forcefully separate your arms apart from one another which *should* rip the duct tape parallel along the axis of the separation.

This is not the only technique for removing duct tape from your arms and legs.

It is more likely that this technique will work if the duct tape is the *only* restraint on you, meaning your kidnappers haven't also handcuffed or flex-cuffed you. It is also more likely that this will work if the duct tape has been applied in a fairly clean and organized and flat pattern. The more that the duct tape folds and rolls etc. the more likely that this technique will require some troubleshooting.

There are some tips and tricks that will help this technique work better. Some things like the position of your hands, the point of impact when you swing your arms apart, using your chest as a wedge, etc.

We have successfully used this technique in our courses on students with bare skin arms and bare skin legs, and we've also had just as much success with students duct taped *over* the sleeves of their long sleeve shirts, and having long pants on where the ankle area of their legs was duct taped.

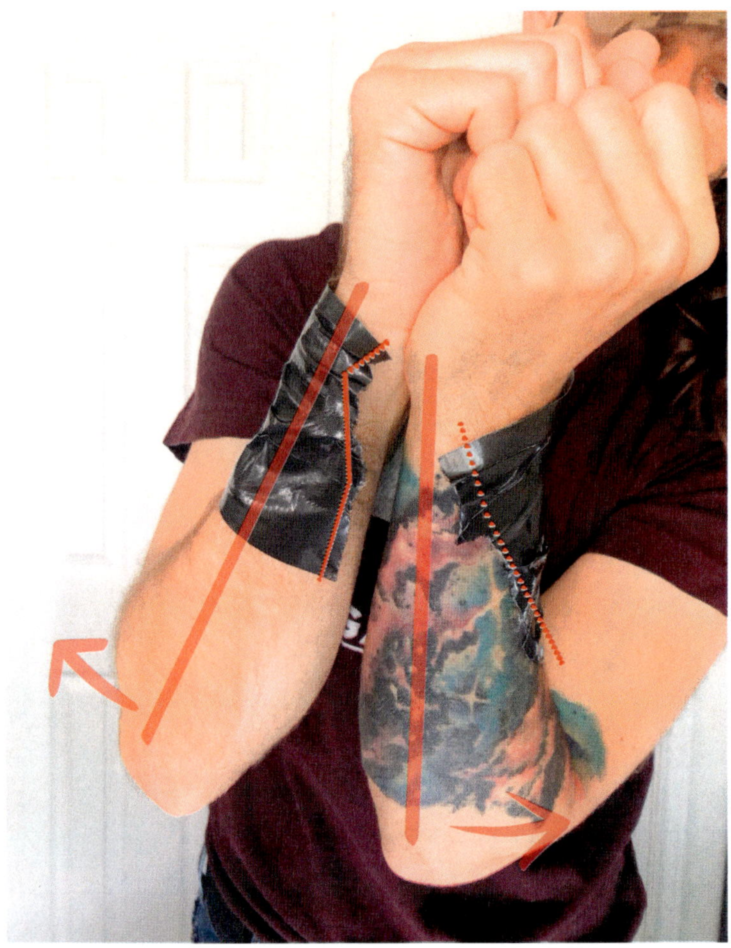

ROPE/CORD/ETC.

You use your teeth, or whatever else is available, to select and pull one strand of the rope and to make it as loose as possible, taking the slack out of all of the rest of the rope. Once that loop is as loose as possible try to slip it over both of your hands, then shimmy your arms to fully distribute all of the slack to all of the rope again. And repeat. The scientific term we teach to our students is the "Loop and Shimmy" technique.

With just this limited crash course above, in over a decade we haven't yet had a student fail to escape a tightly wrapped binding of 550 cord on their first try. This is just to get you started; keep in mind that there are many ways that you can escape from rope as well as other restraints, too.

None of these restraint escape techniques are 100% reliable 100% of the time. There are ways that we can make this restraint more difficult to break out of.

Especially with thinner and tighter restraints you need to be very careful when you plan your personal training program. With tighter thinner restraints it becomes more likely that injury will occur with things like circulation and with injury during escape. Our safety disclaimer at the beginning of this chapter is very important, and this book should not be considered your guide as as "how-to" for restraint escapes. If you're going to practice restraint escapes we strongly recommend you find a trustworthy and professional instructor when you first get started in these escape methods.

Reminder: There are no bonus points awarded for escape-practitioners that practice at home, alone, risking death or injury, while not having any professional escape tools available and without a spotter. It is not fun or sexy to have to call 911 to help you remove your restraints that you've gotten stuck in, alone. Smart is the new sexy. Be smart.

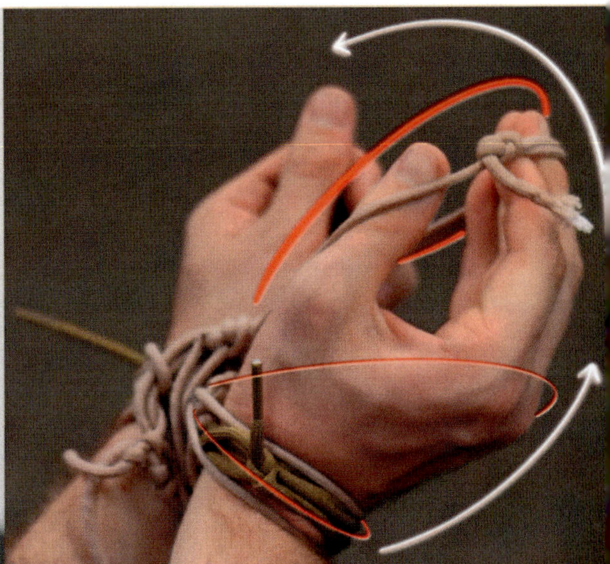

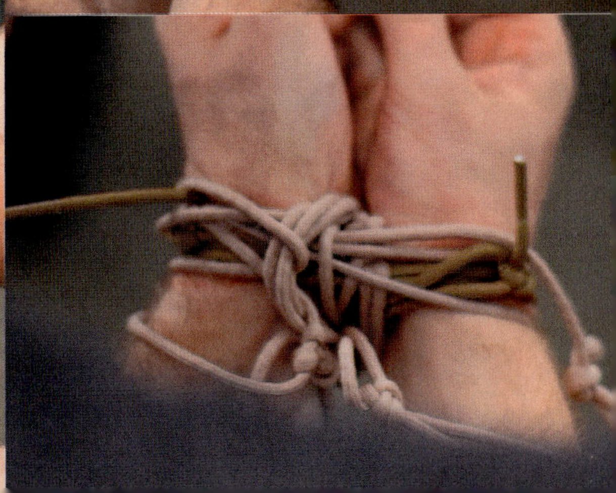

RESTRAINTS CONCLUSION

The study of restraint escapes and the study of overcoming locked obstacles (like picking or bypassing door locks or padlocks etc.) are very similar. They both require the understanding of locking mechanisms, they both use similar language like picking and bypassing and shimming the locking mechanisms, and they both require trouble-shooting and practice and professional commercial grade tools as well as the possibility of using field-made or makeshift tools.

The mental state for when you would practice these two different modes is likely to be very different. One mental state for lock picking is likely to be "I'm intentionally trying to go from common space into a space for an emergency" and the mental space for restraint escapes is likely to be "I am the one locked within this restraint system and my life likely depends upon my ability to defeat this locking system I am bound by."

Lock picking is likely to include questions in the decision making process like "How much time do I have in order to do my research and to find or make tools or to go retrieve tools so that I can unlock this obstacle in the most reasonable manner in order to deliver some type of aid or service" and the thought process for restraint escapes is likely to exist within a completely separate part of the survival-brain with something like "Oh God! They're going to kill me. I'll have to act fast - or wait quietly for the right moment - and I may have to seriously injure myself to get these restraints off or I may have to kill someone, and my life is about to end."

The UTAC staff constantly meets to discuss which tools and techniques we teach and we also talk in depth about how we present the information that we present. With this revision of this first volume of Tactical Lock Picking, we are moving restraint escape techniques slightly lower in the curriculum lineup to take up a more informational position for our readers/students.

We still offer classes on just restraint escapes, and we still include restraint escape techniques in our in-person Tactical Lock Picking courses, but to truly do justice to the world of escaping captivity we decided, for now, to keep restraints as an introductory item and to give them their own space in our course lineup for students that have a desire to expand that part of their skill set.

[Photos below and opposite: Two of our Covert Entry Operationss course students in action, breaking out of their handcuffs after being "captured" for leaving a trace of their entry into the facility upon completing their op. 2022]

CURRICULUM: TARGET ASSESSMENT

Target Assessment is a way more important skill to teach to our audience than any combination of how-to's for tools and techniques on different types of locks. I have a lot of friends, collaborators, and students that send me updates that they've taught their children (some of which are under 10 years old) how to pick open simple padlocks and door locks.

The *skill* of moving around two tools inside a keyway is not complicated at the entry level for most easy locks and even some intermediate level locks.

The skill set of collecting and sorting pertinent information, prioritizing information in a manner that fits your current environmental and mission needs, the skill of knowing which tool and technique to implement on which lock and for how long, and knowing when to stop your attempt to reassess and regroup and possibly to pivot your entry around at a moment's notice to move in a different direction, and how to envelope all of this into a successful and repeatable and organized manner? Extremely valuable.

Good news, bad news. There is not really a wrong answer for how you flow through your decision making matrix on whatever scene you are working. Everything here is a custom approach, and it *has* to be. This is scary for a lot of bureaucrats to hear that "...every part of our SOP's are custom to every individual user..." but there is no other way to teach people decision making other than for them to understand their own strengths and weaknesses and to understand which tools they have on them and which tools are nearby, and to take into account *all* of the data from the scene in front of them.

Present two highly trained and highly qualified CMOE entry specialists with the same exact locked obstacle (the same home, the same business, the same facility etc.). Listen to them talk through their thought process out loud and you should almost certainly see a different flow of information pass through their brains, even if both entry operators use the same tool selection and the same techniques in the same order and they achieve the same entry in roughly the same timeline. Like two droplets of water in a river, even if both droplets start in the same location and end in the same body of water downstream, they will both have a wildly unique path.

THE STEPS

TARGET ASSESSMENT - STEP 1 - MORAL RIGHT for ENTRY (MRE)
TARGET ASSESSMENT - STEP 2 - SPEED

MASTER - **SPIES** - ALWAYS - MAKE - THIEVERY - NOISES - OBSOLETE

Determining your Moral Right for Entry, and the Speed required of you. Steps one and two should happen before the entry begins, they are *almost* always automatic, and should always be reassessed during entry. It's rare that you'll need to make entry somewhere but won't know the why. Imagine how a normal request would sound, "Hey entry specialist, we have this thing going on with this lock and for this and that reason we need you to make entry." You probably now have the reason/morality for the entry. Even if no timeline was given, context clues and tone can usually tell you a lot about the speed. You'll almost always have the ability to ask about MRE and the Speed. I struggle to imagine an example where you wouldn't know the MRE and the required Speed, but it happens.

Easy example: your good friend locks themselves out of their home or their car while you are in the area, and they ask you for help. Morally you are "in the right" to help, and the speed (if we assume this is a simple lock out) is what we would call "administrative" or basically no real rush.

Difficult example: your supervisor knows about your skill set and they order you to achieve a non-destructive entry through a specific door at your workplace. Before or during the entry you realize that the supervisor does not in fact have the authority to open that specific space and when you ask them about their authorization they hem and haw and try to coerce you to continue with the entry and to not tell anyone about it - morally wrong. This coercion happens in real life, ask me how I know. Be prepared to say "no" in a number of unique ways whether it is a bold faced refusal or a creative "ouch my back hurts, I have to go home early, good luck with that door sir/ma'am."

TARGET ASSESSMENT - STEP 3 - ATTACK VECTORS

MASTER - SPIES - **ALWAYS** - MAKE - THEIR - NOISES - OBSOLETE

Identifying all of the possible Attack Vectors. What do you know at this point in time, now, about the locked obstacle that you currently require entry through? Information (or intel) about a locked obstacle or structure can be gained while you're still at home and on your couch. It can be learned over a phone call. Or this could happen while driving to the scene, with someone on scene sending you photos of possible entry points. You can also gather intel while you walk up to the lock. You also have the option to send people around a large home or facility to report back to you with their findings.

This step can be automatic as well and can certainly happen in the background of your brain. It's simply the act of seeing, or feeling, or knowing, and mentally filing any of the possible avenues of attack that your obstacle presents to you. There is however a point in your process when you reassess and start from the beginning where you'll want to *consciously* attempt to list all of your known and possible Attack Vectors when you're troubleshooting and trying to see if you have any better options or any options you may have missed.

When you do take the time to *consciously* attempt to gather all of the possible entry Attack Vectors it can help to jog something loose in your brain. You can also do internet research while on scene, or you can access any of your reference materials (4th Line Gear) to try to discover or learn of any Attack Vectors that may be present but that you may not yet have realized. Taking the time to *consciously* attempt to identify all of the possible Attack Vectors can also help prevent tunnel vision. (The last step in this Target Assessment process loops back to all the other steps: Other Resources, Other Options, Reassessments, and a Reset Breath.)

TARGET ASSESSMENT - STEP 4 - <u>MANUAL</u> UNLOCK CHECK

MASTER - SPIES - ALWAYS - **<u>MAKE</u>** - THEIR - NOISES - OBSOLETE

In many case studies that I've been involved in, the answer is often that the lock is already unlocked and all that is required is that you physically put your hands on the obstacle and attempt to open it. Sometimes it requires a little extra elbow grease - this extra effort is not uncommon, therefore we have tried to beat this entry mantra into the brains of our students:

"Try the lock, take a deep breath, then try again harder."

Additionally:

*"Push and pull the shackle. Twist the knob left and right. Push and pull the door.
Pull down and up on the lever. Etc."*

This is not only for doorknobs and padlocks, it's also windows, latches, chains, gates, etc. While picking open a lock looks super-cool to onlookers, it also looks pretty cool to show up to a scene as the "Entry Specialist" and to twist a doorknob really hard and to easily swing a door open in front of other professionals who believed themselves to be "locked out".

TARGET ASSESSMENT - STEP 5 - <u>TOOL</u> AVAILABILITY

MASTER - SPIES - ALWAYS - MAKE - **<u>THEIR</u>** - NOISES - OBSOLETE

This step has many parts to it, limited only to your creativity, but we'll start you with some common categories.

- What tools do you have on your person, right now?
- What tools do you have stored nearby in a vehicle, or in a bag close by, or in an office?
- What tools do you have off site that you could have delivered to you?
- What tools could you makeshift from your environment?
- What tools could be bought at a store nearby and brought to the scene?

What information (information used as a tool) do you have, or what information exists, that you could look for or research that could assist you with your entry?

All of the tools you purchase and practice with, all of the techniques that you have in your arsenal, and all of the steps of decision making (this Target Assessment process) will improve as you gain more real world experience. **These Target Assessment steps are not designed to give you a rigid process that you must follow in order**. They can certainly help you if you get stuck but they are not to be used in order to "show your work" if you have developed a rapid and likely successful path to entry and if using this format will slow you down. **Break these rules!**

After carrying relatively the same First Line Gear setup for over a decade now, when I realize that I'm locked out of a particular area, and with the moral right to make entry, I can instantly know "I do/don't have the tools in my pocket to attempt an entry for any of the attack vectors that are available to me right now". The power of this instant knowledge really speeds up my decision making process on scene, and for this reason (stated earlier in our gear chapter) I don't change up the tools that are in my Lines of Gear very frequently. If you want to do that, it's up to you but I like the ability of knowing *exactly* which tools are in my lineup because they're *always* in my lineup.

The opposite of a relatively unchanging or slowly changing gear lineup is: buying and collecting and reorganizing tools and then having to spend time looking through your Second Line Bag on a discovery quest to see *if* you have a tool for a particular Attack Vector. Not "wrong" in and of itself, but imagine if we again use the lens of a high-speed military pararescue operator who just collected a casualty from a combat zone...

> "Okay I've got the casualty onboard the helicopter now. I have just observed a particular type of injury. Well... let me rifle through all of the different pouches and bags of medical equipment that I have onboard to see if I have a tool for this problem" ...wow. So, that... versus the equivalent of immediately selecting the best possible tools from a drop-down menu and taking immediate problem solving action should a particular tool be known to be in your kit, or not.

As your tool selection grows and changes, as your skill level changes, as you start racking up more case studies, this process will grow to match your progress.

TARGET ASSESSMENT - STEP 6 - <u>NOW</u> OR LATER

MASTER - SPIES - ALWAYS - MAKE - THEIR - **<u>NOISES</u>** - OBSOLETE

This step is where the entry specialist computes some simple math (don't get scared) and then asks the question "Should I start putting tool to lock now, or should I continue trying to gather more information before starting my physical entry attempt?"

For the math, it's not necessarily actual numbers. You can work in simple mathematical *concepts* or in rough "back of the envelope" whole numbers.

If you are locked in a state park like our earlier case study "Locked In", the math for this step might look something like the following for you, if you were assessing your own exit out of the state park: "Ugh... I could start counting numbers on this combination padlock and it might be really fast or really slow to find the right code. But I can literally *see* a hardware store from here where I could go buy some bolt cutters and walk back to the park and cut this padlock (if that's reasonable for this imaginary case study)". That math wasn't so scary, was it?

Another example might be: "This facility is about the size of a football field and the owner, a friend of mine, sent his facility keys home with an employee and now we're locked out of this training facility. I see that he has a Smartkey lock on the front door and unfortunately I don't have my Smartkey camera decoder or my key cutting equipment which would be my first choice. I do have my UTAC Smartkey Field Tryout Set with me but that could take anywhere from 1 minute to 20 minutes. Before I start that 1-20 minute entry attempt let me take 5-10 minutes to walk around this football field sized facility to see if there's a door latch that I'd rather slip open instead." Boom. Conceptual math complete.

In this Target Assessment step, if you answer to yourself "Now" that you're starting your entry attempt, you should implement a physical or at least a mental timer for your "Line in the Sand" (LITS) guideline. This is how long you'll spend on a particular technique before you force yourself to reassess and then to pivot to another Attack Vector or even to *intentionally* continue with the same method that you're currently working on.

If your answer to yourself is that "Later" is when you'll try the current Attack Vectors that you have at hand, then go continue looking for more Attack Vectors or more resources or more information.

And remember, just like the famed "OODA Loop" this Target Assessment procedure does not have to be consciously spoken out loud in the heat of operations. You don't have to say "I have this Moral Right for Entry. I know my Speed. I have seen several Attack Vectors. I will now Manually attempt to check this lock." You're allowed, and encouraged, to walk right up to the front door and twist the handle.

TARGET ASSESSMENT - STEP 7 - <u>OTHER</u> RESOURCES / REASSESS / RESET BREATH

MASTER - SPIES - ALWAYS - MAKE - THEIR - NOISES - **<u>OBSOLETE</u>**

Once you have struggled with an entry, once none of your currently known Attack Vectors are presenting you with a reasonable entry option, just stop... and just breathe.

***Breathe*.**

Going back to the beginning, all the way back including "Do I have the Moral Right for Entry, and what is my Speed requirement" is always an option. And the answer might be that "No," that you should just stop your entry or to entirely pivot to a completely different entry or non-entry solution.

Going back and walking through everything you've already identified and tried is also a really good option. I have used this to save myself on multiple lock-outs. One of my favorites is when I tried and tried and tried to get a latch slip to work for a bathroom door and only after exhaustion I performed a walkthrough and reminded myself "Did I try to reset that deadlatch mechanism again" and immediately corrected my actions and gained successful entry.

Who else can I call? What can I research online? Let me step back to see if I can collect some new information or look closer to see if I've missed any small details. Go nuts. Be creative. Let your brain relax for a second. The answer might be to continue struggling with a particular technique, the answer might be to call for help. It might be to keep looking for other entry options. This step is where you gather with the other people that are helping you or where you sit with the imaginary council in your head to talk yourself through other creative options.

[Photo: Terry (right) of CDI Tactical, one of my TLP instructors, and myself (behind the camera) during a military CMOE training contract in Europe - Jan 2021]

YOUR ENVIRONMENT SPEAKS TO YOU

Your environment speaks to you...let it. And listen.

Keep an open mind and learn how to listen to your environment. The more tools you learn how to use, the more techniques you learn how to apply, the more locks that you successfully open, the more locks that you *fail* to open, the more you will see things in your environment that draw connections within your skill set.

Seeing a huge gap under a door that has a handle on the front, probably has a handle on the back and can probably be opened with a UDT. Your environment may tell you "Try the UDT first!" [Photo above]

You may pull up to a locked door in your vehicle and from within your vehicle you may see the reflection of the single latch within the wide open gap between the door and the frame. Your environment is telling you "Forget the lock picks, just use a pocket knife and this entry is likely to take about six seconds."

Your environment might tell you "There's no need to do any Target Assessment at all, just start with this Attack Vector RIGHT HERE. If it doesn't work immediately, then you can go start an assessment."

Your environment might tell you "Other people have bypassed this lock before, many times, in this one particular way," by seeing common evidence of a bypass that is left on the obstacle.

A door might have multiple expensive and correctly installed deadbolts on one side, but it might be a pull-door and the hinges might cost pennies and might have their hinge pins loosely hanging out of the top of the hinge. This door is ready to have the hinge pins pulled out by hand and will allow you to pull the entire door off the frame, ignoring any locks.

When you're stuck knee-deep on a non-destructive entry operation, when you stop to take that Reset Breath and to reassess everything you've already tried and you're looking for Other Resources, let your environment speak to you.

TARGET ASSESSMENT OFF SCENE

One of the greatest things about this Tactical Lock Picking skill set is that it is "highly transferable".

You can be a 20 year veteran of special operations and you can teach most of your entry techniques to your 10 year old child; transferable.

You can be a bodybuilder and a lock picker and you can teach all of your lock picking techniques to the sweet elderly lady that lives next door to you and she has the potential to be just as good as you are at Tactical Lock Picking even without all the muscles; transferable.

You can give someone a crash course with a particular tool and technique and they can often achieve an entry (on low security locks) almost immediately; transferable.

You can get a phone call from a loved one from across the country and you can verbally walk them through how to find a tool in their environment and how to apply the tool in the right conditions even when they don't have professional tools or training and even when you're not on scene with them; transferable.

You can recon/research a lock or facility while sitting at home in your underwear; transferable.

Before you arrive at the lock-out, you can have someone one scene call you and explain what they see, or they can send you images or videos of the door and locks on site while you're still en route to the location; transferable.

You can even gather information about the lock-out, while enroute, from someone else that is also off scene but is familiar with the locks or the facility in question; transferable.

This skill set doesn't exist in a vacuum. And results in the field are not scored with a grade, only with "Pass/Fail".

[Photo below: Sky and Mai of CloakedEntryCo.com during a Tactical Lock Picking collaboration course - 2022]

TARGET ASSESSMENT CONCLUSION

If we took an MMA fighter who was really good at punching combinations, really good at big-strong-fast-powerful kicks, really good at staying on their toes, but has never ever stepped into the ring to fight an opponent in a ring-fight and has also never even stepped into a practice-ring for sparring... he's just a hobbyist. There is nothing wrong with being a hobbyist but there *is* something wrong with being a hobbyist and *thinking* that you are a fully prepared and fully realized ring-fighter.

Being really good with your pistol and your rifle at the local indoor shooting range. Being really good at drawing from the holster. Being really good at double-taps. Having great speed and accuracy while shooting your paper target, that's an important (and small) aspect of self-defense shooting. But if you think that maxing out your indoor range experience without ever taking a CQC or force-on-force class makes you combat ready, then that is a lie that you've told yourself.

The value of the ring fighter isn't just how good they are at punching, it's how they manage the entirety of the fight and how good they are at delivering a winning result.

The value of the CQC operator is not measured in speed-shooting competitions. Their value is in keeping innocent bystanders, team members, and even suspects as safe as possible while also obtaining maximum mission completion.

For Tactical Lock Picking, for the Covert Entry practitioner, the value is not in how difficult of a padlock the operator can pick open with a hook and a tension wrench. The value of the TLP operator is their skill "in the ring," their skill "in the shoot house," or more accurately "on the battlefield." The measure of a competent TLP practitioner is in how they navigate the *operation* and in their ability to choose the best possible path to mission success.

DON'T BE AFRAID TO FAIL. FAILURE IS A NATURAL PART OF THE LEARNING PROCESS.

GO TRY. GO FAIL. GO LEARN. THEN GO GET BETTER.

YOUR FAILURES WILL HAVE MORE OF A POSITIVE IMPACT ON YOUR FUTURE TARGET ASSESSMENTS THEN ANY OF YOUR SUCCESSES.

CURRICULUM: THE FIELD IS NOT YOUR COUCH

There are "tools" that you will have to use in the field that simply don't exist on your couch.

While picking a lock on your couch you can: wear what you want, eat and drink, sit back, set the lighting, never drop a tool too far from the lock, have access to any tool that you own, take your time, etc. etc. etc.

This list is not all inclusive but here are some "tools" that you will want to be aware of in the field so that you can have the best possible edge on your entry likelihood percentage.

ANCHORING

Anchoring: If your environment allows, always try to manage how close and how steady you can keep the locked obstacle and your own hands and body during your technique application.

A word on vices. A lot of people, especially in the Locksport world, practice their craft with a padlock or simply the core of a lock mounted onto a tabletop vice.

This is great. I support this. In the same way that I support people practicing "Sport Biting Competitions" with their dogs I support people using a desk vice to practice their entry skills. What I don't support is people *thinking* that their dog engaging in a bite against a well trained practice "decoy" in a bite suit during a rehearsed demo with a high level of accuracy and repeatability is the same as how your dog would handle a real attacker (or attackers) with weapons and with determination to cause you harm.

In the Tactical Lock Picking space here, this is the same as a practitioner thinking that "I can pick this difficult padlock repeatedly and successfully (in a vice)" and then them thinking that they will have the same repeatability and the same success rating in the field with that same difficulty of lock when it is cold, and wet, and dark, and the padlock is not mounted in a vice but it is hooked into a chain and stuck on the far side of a fence that they have to reach through and when they visually *can't even see the lock* but have to feel their way into it with their tools.

I do support people increasing their skill level by isolating movement within a lock by using vices. For the field operator, I only ask that they understand intentionally that it is a tool for isolating a variable and it is not a tool that should give you a direct translation of your skill level outside of the vice.

Enter the Tactical Lock Picker's tool of "Anchoring" that doesn't exist on the couch.

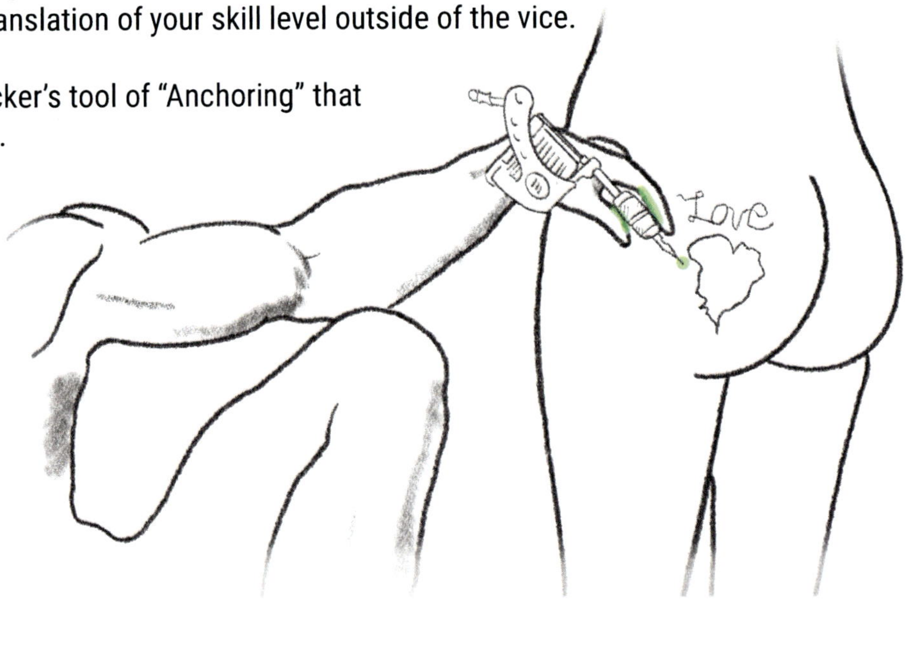

I hope you're not sick of analogies yet: Imagine a tattoo artist standing at arm's length from the customer and extending their arm, with the tattoo gun at the end, and them *not* resting the palm of their hand on your body and them not using a support hand. Just basically playing "Pin the tail on the donkey" but with a tattoo gun. Is that the type of tattoo that you want on your skin? That is clearly an imaginary scenario because what real tattoo artists do is they "anchor" their hands onto the customer's skin so that they are using only the smallest amount of movement possible and they are eliminating any excess movement out of themselves while they create your art. We hunt for the same thing in the field.

If I'm picking a lock on a chain that is holding two gates together, not only will I anchor both hands around the padlock but I'll *also* anchor may hands *and* the padlock to my leg or to my hip, sandwiching the padlock into a position with as little movement as possible so that I can focus my attention on the movement of the pins and the tension within the plug, instead of trying to also manage all the movement of the swinging padlock at arms length.

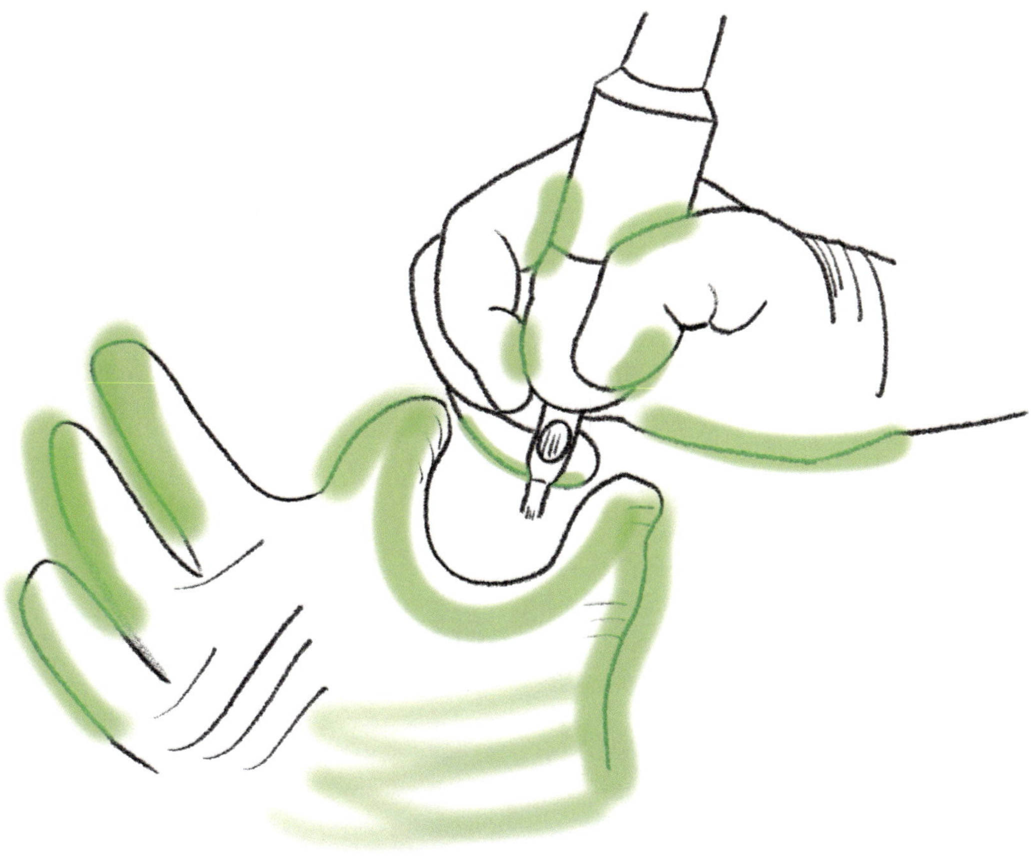

[Photos opposite and above: A tattoo artist doesn't do intricate artwork at arms length, instead they **plant their hands firmly** around their needle area, **getting as much of their skin and their musculature frame in a locked-in and anchored position**. When dealing with a lock in the field, you should try (if possible) to not only get the lock into a more accesible and secure position but also to put your hands and your body in the most comfortable position that your obstacle in the field will allow.]

POSITION OF COMFORT

Position of Comfort: In addition to anchoring, when possible, TLP operators should always seek to move not only the lock but also their bodies and also the entire obstacle itself in order to focus as much on the locked obstacle as possible while removing any position-related variables that could add complexity to the entry attempt.

Very closely related to anchoring, position of comfort is a tool that simply doesn't exist as an option on your couch. This is because on your couch a padlock is simply held in your hands. In the field, a lock that you'll need to access will always be mounted onto an obstacle.

On your couch, of course you'll be "comfortable," but in the field is the only place that you'll have access to *the act of* "seeking a position of comfort".

Be creative, the sky's the limit. Sometimes kneeling at an obstacle will be the most comfortable position for you at the time. Sometimes it's a rotation of your body left or right. Sometimes, for locked overhead obstacles, a ladder might be your best option to increase your position of comfort and to increase your chances of a successful entry. Tactical Lock Picking is about the application of your craft, not just about picking locks.

Take a large plastic gear case that is padlocked shut without the key. The particular cases that are the size of small coffee tables. If the case is sitting flat on the floor then a proper padlock would be hanging onto the case lid and the keyway of the padlock would be hanging down and facing the floor.

[Photo opposite-top]
At a beginner's level approach (this book is dedicated to a *systemized approach*), the lock won't be rotated, the case won't be moved, and the operater won't adjust their body for a position of comfort. The lock is treated "as is."

[Photo opposite, middle left and right]
For an intermediate level approach, most intermediate entry operators will adjust their body in order to access the keyway better, or they will adjust the padlock so that it's easier to access, and it's also possible that an intermediate TLP operator will do both at the same time.

[Photo opposite-bottom]
Experienced operators will likely scale out a little bit more and will flip and rotate the case, they'll rotate and adjust the padlock if need be, and they'll put their body and their hands in the most comfortable position in order to effect their entry. The concept here is that the advanced operator will adjust *as many* variables as possible to remove distractions, to increase comfort, and to support the main focus which is the mission.

Let's steal another analogy from the shooting world. Who out there would hear about a world renowned sniper that just made a 1-mile long headshot and then would complain "Yeah, but the sniper made sure that he was really comfortable first so that's basically cheating." Sometimes being able to move and shoot (or pick a lock at arms length through a fence) is a mark of an expert. Other times the ability to really ground yourself and settle in for a mile long shot (or really adjusting the lock, and your body, and finding the best fitting tension wrench possible, and breathing slow) is also a mark of a professional.

Remember to think *outside the box* (please clap, humor is not our strong point but we try very hard). In a competition, rules are followed and points are awareded and taken away. During real life emergencies you should assess every possible reasonable edge that you can take in order to help increase your likelihood of mission success.

BEGINNER APPROACH:
attacking the lock and the keyway "as is".

INTERMEDIATE APPROACH:
Adjusting your body only, or the lock only.

ADVANCED APPROACH:
Adjusting as many variables as possible (body, lock, object, and more) so that focus is not diminished.

HAND'S JOBS

Hand's Jobs: "Which specific jobs are each of my hands responsible for during an entry attempt?"

On the couch, the "*desire* to stretch your skill set" rests solely on you. This could be anywhere from your brain wanting to stretch a little, a lot, or not at all. And that's fine. But in the field, there is a "*requirement* to stretch your skill set" that is always 100% of whatever the obstacle demands of you until you make an entry or until your non-destructive efforts are exhausted or no longer the best option.

Get comfortable multi-tasking with different parts of your hands and get comfortable using different hands for different jobs, a lot.

I think what causes a lot of people to shy away from practicing an ambidextrous approach in the tactical world is the liability or risk. If you're practicing shooting (yes I know a lot of people do practice with both hands) there is a large segment of people that may consciously or even subconsciously think that if they switch hands and make a mistake, the liability and the damage is elevated. This makes "failure during learning" a lot more scary.

In lock picking, the risk involved in making a mistake by switching tools into different hands is ZERO. And one of the very specific framing items of Tactical Lock Picking being the framing for "in the field / emergencies / operations" means that the emergency presents itself to you and you don't have a choice but to troubleshoot unique problems that exist in real life. In the field you're not allowed to say "Excuse me, instructor, but this lock has been installed too closely to the frame of this door and I can't fit my tension wrench into the lock in the normal fashion that I'm comfortable with, can I have a new training doorknob please?" Your only option is to say "If I'm going to gain non-destructive entry, and if I'm going to attack this keyway, I've got to be creative with which tools go where and which hands operate which tools in which direction."

Practice using your left hand as your tension hand and your right hand as your raking and picking hand. Then reverse and use the hands for the opposite purposes. Use your pointer finger for tension, then try again by switching your hand around and using your three support fingers to manage tension, then try using your thumb as your tension manager [Photos below]. Try both hands that way. There have been a few very unique field obstacles I've come up against where I've had to use just **one hand** to be my raking hand and my tension hand due to limited space.

[Photo above: Standard pin tumbler low-security padlock with a "flat" tension wrench in the bottom of the keyway, the tail of the tension wrench pointing up towards the "North" or "12 o'clock" position.]

[Photo above: Standard pin tumbler low-security padlock with a "flat" tension wrench in the bottom of the keyway, the tail of the tension wrench pointing down towards the "South" or "6 o'clock" position.]

[Photo above: Standard pin tumbler low-security padlock with a "bent" tension wrench in the bottom of the keyway, the tail of the tension wrench pointing left towards the "West" or "9 o'clock" position. *The direction does not need to be exact for any of these.]

[Photo above: Standard pin tumbler low-security padlock with a "bent" tension wrench in the bottom of the keyway, the tail of the tension wrench pointing right towards the "East" or "3 o'clock" position.]

For field operations you'll want to be comfortable using top of keyway tension and bottom of keyway tension, and you'll want to be familiar with the tail of your tension wrench sticking out of the lock in all four of the cardinal directions. When you first start learning, you'll probably find a hand position that works for you but if you have any intention of picking locks *during emergencies* then you'll want to start adding different hand positions into your training regimen immediately. This experimenting in hand-usage preparedness will help you to solve complex and unexpected problems in the field with much more speed and confidence.

GRAVITY

On your couch, sure you'll drop tools and have to search for them. Tools can and will go springing and flying all over the place. The difference is the learning stimulus. On the couch there's no real learning-pain if you have to slide down onto your hands and knees and brush your hands over your carpeted floor to look for a tiny tool. In the field, possibly at the top flight of an outdoor fire escape, the learning-pain stimulus for dropping the sole specific tool that you have for this type of lock is a hell of a different stimulus.

On the couch, there's no need to adjust the way you store or retrieve or hold or use your tools because there is minimum if any real "stimulus for change". In the field, the first time you have to walk several flights of steps (or even one flight of steps for some people) it is an eye opener and it demands a change in action in order to avoid that same pain in the future; you must change the way you operate in fear of the same problem occurring again.

This "tool" that we call gravity is not just about dropping tools. Consider the overlap in this book, and in life, anchoring and positioning can have an impact on fatigue; use gravity to position your hands and the lock and your body in the most comfortable position that your operation and environment allow. Consider the need to lift a heavy tool up to a locked obstacle or rather taking the locked obstacle down to floor level before you use the heavy tool.

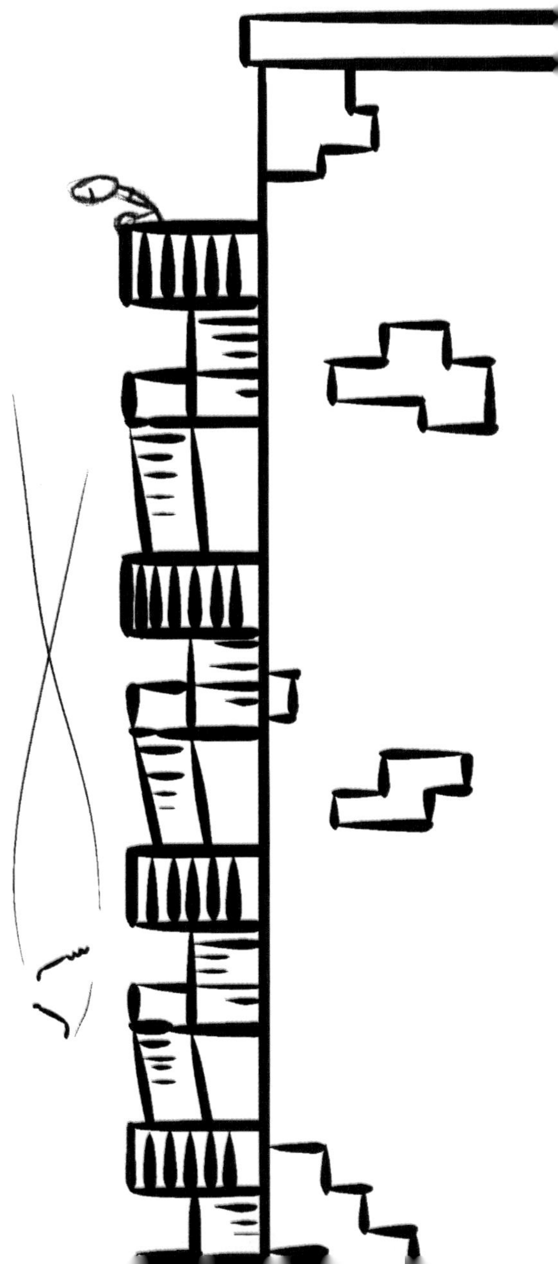

Even the act of getting a flexible door latch shim to slide through the door frame of a sticky door might be easier if you start the shim high and use your body weight to pull it down the door frame, rather than starting low and using only your muscles to pull the door shim up to the latch.

If we're going to use all of our advantages that we can during an emergency to achieve an entry, we should use all the tools and all the forces at work when we're able to. Gravity...is one hell of a force. A hard lesson for you to learn if you are unprepared: pulling the hinge pins out when taking a door off of the hinges for an entry, if you take the top pin out and then the middle pin and then you crouch down to remove the bottom hinge pin... watch your head.

When you struggle, and when you take that Reset Breath and you consider your circumstances, try to remember to *always* use gravity to your advantage, whenever possible.

LIGHTING/VISION

If your goal is to use this skill set during emergencies then your practice should mirror the intended application of your craft. Again, there is no need to find solutions to enhance your lighting or your ability to see when you're practicing on your couch while binge watching your favorite show. This is not a value judgment, picking on the couch *does* have benefits. We just want you to be well informed that spending 8 hours a week picking locks at your desk does not equal 8 hours of "solving problems in the field".

There will be times you will have to identify a lock without seeing it, you'll have to get your tools inserted into the lock without vision, and you'll have to troubleshoot without any lights and at arm's length, upside down, etc. etc. etc. Not every time, but when you *need* the ability to increase your vision capability or when you have the need to be practiced in troubleshooting without sight - you'll want to have had experience doing that before and not in the heat of the moment.

Pack a headlamp in your Second Line Gear bag. Turn on the lights. Have someone hold a flashlight. Use a mirror. Prop your phone up on the far side of a gated obstacle and watch live through your selfie-camera on your phone screen. Slide that phone halfway under a door while attempting an Under Door Tool entry [Photo background]. Hold a video call with a friend who is standing on the far side of a structure who can see inside the structure.

While some lock picking courses exist in the academic realm (teaching students how to use tools and techniques on certain locks), *we* exist to teach professional and self-supported operators how to solve locked obstacles (while also teaching the nuance of tool and technique applications).

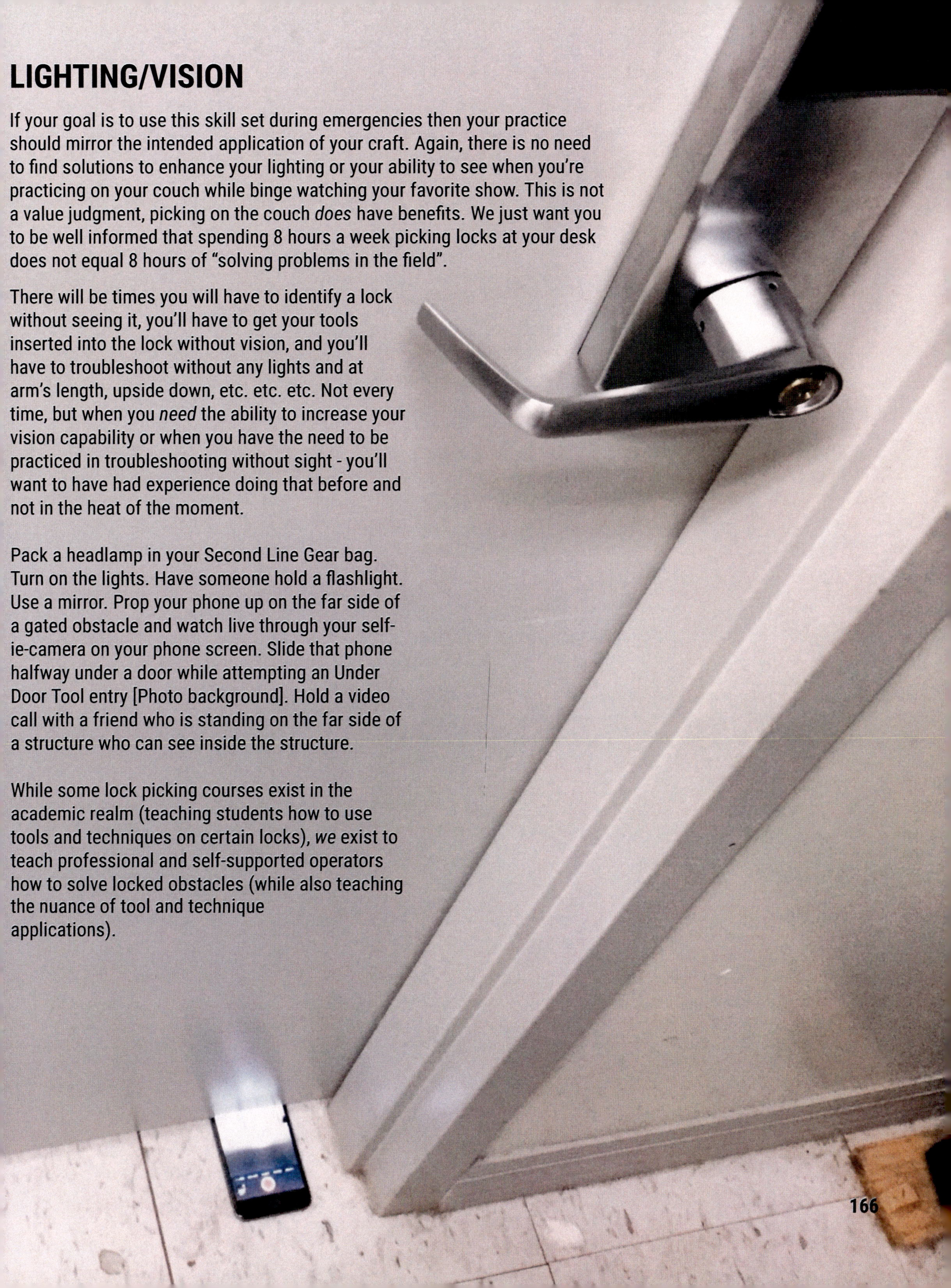

FIELD/COUCH CONCLUSION

Emergencies and operations do not care about which tool and which technique you apply to which locks, if any. They only care about a resolution.

CURRICULUM: FAILURES

This chapter (just like all the others) could be its own whole book.

Imagine you are requested to make entry using your Tactical Lock Picking skill set, non-destructive and/or a CMOE entry; imagine you carry out the entry and every tool and technique works exactly as planned and no troubleshooting is required. What have you learned? The only *possible* thing that you can objectively add to your learning pool is that "yes, this worked".

Now, what if you had to struggle during the operation and had to switch a few times from one tool/technique to another two or three tool/technique options but you eventually did successfully achieve entry? What would you be able to objectively add to your learning pool now? "This didn't work, possibly because of X, Y, and Z, but this other thing did work," and more, much more.

You still have to go out and fail and learn, but we'll list some common failures here below:

- Often tools will fail/break/be useless upon arrival from their manufacturer. Sometimes it's a manufacturing flaw, sometimes a design flaw, sometimes it's a lie by omission where the retailer doesn't tell you that the tool *doesn't work* on specific makes and models or generations of locks. Sometimes tools just don't work.

- Tools break.

- Batteries fail.

- Some tools require maintenance and it's not always immediately clear when that's required, and some tools fall out of calibration.

- Some tools, by design or by accident, commonly become "irretrievable" meaning that they become temporarily (or worse, permanently) stuck inside of locks.

- Sometimes you just have a plain old bad day and can't get an otherwise working tool and correct technique to open up even a simple lock. It happens to me and I've been a SME in this field for over a decade, and it still routinely happens to friends of mine that are ranked world champions at speed lock-picking. In the entry field, if your instructor ever tells you that any technique out there will work 100% of the time, or if they fail to mention to you that all techniques are percentage based at their likelihood of success - they are fucking liars.

- Sometimes locks are weathered, worn, damaged, or sometimes you're dealing with a lock that is new from the factory and they have a different design internal that prevents the type of attack that you otherwise expected would work.

- By tracking which types of locks your skill is unable to defeat, failures can be a great guide to knowing where *not* to waste your time during an entry; and this is not a bad thing. It's a *great* thing to know how to use your time and resources wisely.

This is by no means an exhaustive list but these are some starters that should *actively* help you shape your skill set - and that's the whole purpose of this book, *helping you build a skill set*, not just showing you a how-to for a tool and a technique for a particular lock.

CURRICULUM: DEFENDING AGAINST CMOE ATTACKS

Beating a dead horse here but, this chapter could also be its own volume within our Tactical Lock Picking series.

The first thing you need to know when you are taking a look at your own security system for the first time, really opening your eyes to just how insanely easy it is to overcome the illusion of security, is that: **THERE IS NO SUCH THING AS TOTAL SECURITY**. And that's okay.

The second thing you should know is that security is like an onion, or like an ogre; **it has layers**. The best security assessment question is "How can I design some security layers to better fit this system?"

Next up, **almost all security is designed to buy you time** and not to "completely prohibit access". Even the toughest of super-secret high-level government bunkers can be penetrated with the right use of resources.

Security is always a balancing act between effectiveness and efficiency. If you are about to add a security layer into your life and that specific security function makes your life more of a hassle, then it's probably worth reconsidering. The absolute best security devices out there are the ones that *blend in* with your life rather than making you go out of your way in order to access whatever it is that you want to access.

Imagine coming home from work every day and having to go through a 17-lock dance at the front door,
and then having to do a retinal scan, then having to sing the entirety of Con te Partiró for voice analysis,
and submitting fingerprints for all 10 digits as well as palm prints for each hand.
What happens when the novelty wears off?
The more frequently you need to go in and out of your home,
the more likely you are to start using less and less of that intensive system.
It will eventually balance itself out.

Match your security posture to the type of threat you are expecting. If you're playing soccer (or European "football"), if everyone's running around kicking things at lower-body height then you're probably much more likely to get kicked in the balls than in the head (for the men out there anyways). Wearing a protective groin cover is a better use of equipment than wearing a helmet on your head. For your home, it's probably a way better use of your resources to increase the strength of your door frame hardware to prevent your door from getting kicked in than it is to worry about buying the right high-security keyway that is "pick proof".

If you are a common American living within a common budget in a common type of residential home,
and Jason Bourne wants to sneak into your home and murder you while you sleep,
there's no amount of preparation that you could reasonably take in order to prevent that.

Some people may have a specific reason (stalker, status-based kidnapping for ransom, etc.) to have an exhaustive and in-depth security posture. For most people, you'll get a much better return on investment **spending 20%** of your original security budget on solutions that will **prevent 80%** of the likely types of break-ins.

For the most part, most locks and most locking systems are actually pretty good at doing their *intended* job (which is: keeping *most* of the wrong type of people out). For us, it is more important that you understand the *design* of security systems, than that we give you a list of *"buy these security devices"*.

THANK YOU

Speaking for the UTAC team, I sincerely hope that the thing that shines through the most in this published work is that **we have a passion for this unique and useful craft** and that we are anything but a boring and sterile academic teaching platform.

It means the world to us that you have trusted UTAC with your training time and training budget by purchasing and reading this book.

Our goal was to revise this written work and to make sure that every single page was packed full of value to help you design and improve your craft and that it keeps within our principles and our focused framing that we try to deliver with all of our projects.

Sincerely, thank you for reading our work,

Pat Watson
Founder - Uncensored Tactical
Email: Pat@utac.io
2023

RESOURCES:

- Come join us at one of our **in-person courses** that we offer on Tactical Lock Picking, Field Key Generation, Vehicle Entries & Restraint Escapes, Covert Entry Operations, and more at https://UTAC.io/Courses.

- **Enroll in "UTAC GROUP."** It's our small-group private underground college where we have 24/7 access to the group via our Discord platform. We try to meet at least twice a month for live video lessons primarily on Tactical Lock Picking. We also have hangout events, we spread around gifts and discounts and we talk about behind the scenes updates. We only open our enrollment twice a year and seats *always* sell out fast! Gain access through Patreon.com/UTAC at our "UTAC COLLEGE" tier.

Access our online video training course for hours of instructional videos that you can watch at your own pace, and where you can hear us talk about more nuance and see us demonstrate tools/ techniques in action. Currently hosted through Teachable. Normally ~~$450~~, available at a *huge discount* for readers of this book at $99! QR code [right] will get you acces to the discounted reader rate or this link: "Reader Discount"

Expand your physical bookshelf with our works!

- Expand this volume with "*Tactical Lock Picking Vol 2. Field Key Generation*."
- Grab an issue of "Covert Entry Magazine", which is a limited run, only available during pre-sales.
- Add our "Brute Force Notebook" to your entry bag to help with decoding mechanical locks.
- And stay tuned for more of our written works, that are on the way!

https://UTAC.io/books.

Grab some tools from our web store to go along with your training and your entry application. Visit UncensoredTactical.com.

And on a budget you can always follow us on our **social** platforms and you can listen to hundreds of hours of audio **podcasts** where we go *way* in depth on the basics as well as the advanced nuance of Tactical Lock Picking and Key Generation, and more.

*Uncensored Tactical is now offering our services to help others in the tactical space to bring their course concepts to life in exciting and engaging written works. If you want us to help you to create a book like this for your brand, please get in touch. Email: Pat@utac.io

REVISIONS

"Tactical Lock Picking - A Systemized Approach to Overcoming Locked Obstacles During Emergencies" **was first published in 2020**.

This revision (the 2nd edition of this title) "Tactical Lock Picking Vol 1. - Foundations" has been modernized, upgraded, and **re-published in 2023**.

For this 2nd edition of "Tactical Lock Picking Vol 1. - Foundations":

- We've taken out a few case studies. We've added in several subchapters. We've taken out a lot of the police-centric content. And we've cleaned up the swearing (not entirely, but quite a bit).
- We've shortened and condensed some of the written content while wildly expanding the information delivery capability by adding higher quality photos, blending graphics and artwork with photos, and using a higher quality print company and paper.
- We've added a lot of the "nuance" that we've been honing over the past few years.
- We gave the gear chapter a huge facelift.
- Our four-digit code chapter got an overhaul and upgrade as well, using our in-house artwork to help give a visual aid to some of the written content that by itself could be dry or difficult to conceptualize.
- We redesigned the cover to fit the format of "part of a volume" instead of appearing as a standalone book.
- We added a few more little lessons learned here and there.
- And more.

-END-

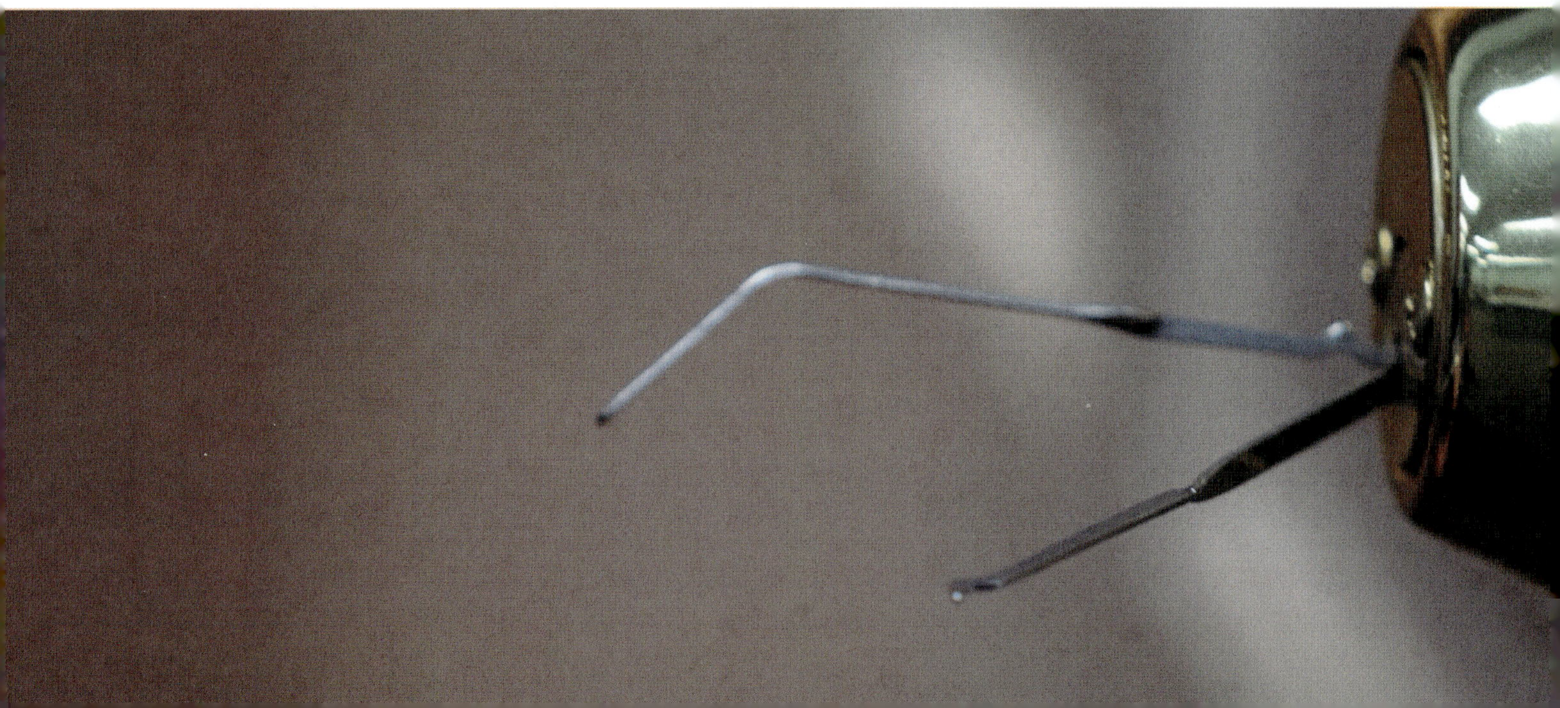

[INTENTIONALLY BLANK]

[INTENTIONALLY BLANK]

[INTENTIONALLY BLANK]

Made in the USA
Coppell, TX
13 September 2026

89920726R00112